GROWER'S GUIDE
Jean-Martin Fortier
FROM THE MARKET GARDENER

The Well-Planned
Vegetable Garden

A Grower's Guide

TRANSLATED BY LAURIE BENNETT
EDITED BY PIERRE NESSMANN
WITH THE COLLABORATION OF TOM LE JARDINIER
ILLUSTRATIONS BY FLORE AVRAM

new society
PUBLISHERS
www.newsociety.com

Copyright © 2026 by Jean-Martin Fortier. All rights reserved.
Translated by Laurie Bennett. Cover design by Diane McIntosh.
Printed in Canada. First printing January, 2026.

© Delachaux et Niestlé, Paris, 2024
First published in France under the title: *Une année de legumes, Les guides du jardinier-maraîcher*, Jean-Martin Fortier, Flore Avram.

The author and publisher disclaim all responsibility for any liability, loss, or risk that may be associated with the application of any of the contents of this book.

Inquiries regarding requests to reprint all or part of *The Well-Planned Vegetable Garden* should be addressed to New Society Publishers at the address below. To order directly from the publishers, please email info@newsociety.com or order online at www.newsociety.com.

Any other inquiries can be directed by mail to:
New Society Publishers P.O. Box 189, Gabriola Island, BC
V0R 1X0, Canada

New Society Publishers is EU Compliant.
See newsociety.com for more information.

LIBRARY AND ARCHIVES CANADA CATALOGUING IN PUBLICATION
Title: The well-planned vegetable garden : a grower's guide / Jean-Martin Fortier;
  translated by Laurie Bennett;
  edited by Pierre Nessmann; illustrations by Flore Avram.
Other titles: Année de légumes. English
Names: Fortier, Jean-Martin, author | Bennett, Laurie, translator. |
  Nessmann, Pierre, editor | Avram, Flore, illustrator.
Description: Series statement: Grower's guides from the market gardener; |
  Translation of: Une année de légumes.
Identifiers: Canadiana (print) 20250278901 | Canadiana (ebook)
  2025027891X | ISBN 9781774060209 (softcover) |
  ISBN 9781550928136 (PDF) | ISBN 9781771424097 (EPUB)
Subjects: LCSH: Vegetable gardening—Planning. | LCSH: Vegetable gardening. |
  LCSH: Organic farming.
Classification: LCC SB321 .F6813 2026 | DDC 635—dc23

Funded by the Government of Canada | Financé par le gouvernement du Canada | Canadä

New Society Publishers' mission is to publish books that contribute in fundamental ways to building an ecologically sustainable and just society, and to do so with the least possible impact on the environment, in a manner that models this vision.

Creating a future where humans live in harmony with nature and with each other

Founded by Jean-Martin Fortier, the Market Gardener Institute is committed to inspiring and supporting new organic growers at every stage of their journey. Our mission is to equip them with the essential technical skills needed to thrive in their vital agricultural work.

Our vision is to multiply the number of organic, regenerative farms around the world and create a future where humans live in harmony with nature and each other.

www.themarketgardener.com

Presenting the collection
Grower's Guides from the Market Gardener

Hi!

I am delighted to bring you this new collection of practical guides. The advice you'll find in these books is based on working methods I developed on my own microfarm and refined over the last two decades. While plenty of these concepts are not new and were passed on to me by different mentors through the years, many other ideas stem from my own farming experience. I am sure you'll come across a number of tips and tricks that are innovative, proven, and easy to implement.

Whether you are a home gardener, hobby farmer, new market gardener, or an experienced farmer looking to transition to more intensive growing on smaller plots, you will find everything you need to take your horticultural practices even further.

Wishing you success and happiness in your agricultural adventures!

Jean-Martin Fortier, market gardener in Saint-Armand, Quebec

Contents

Introduction: A Few Words About My Background	1
What Is the Market Gardener Method?	4
Preface: Planning, the Key to Success	8
Why Make a Plan?	9

USING THE MARKET GARDENER METHOD TO CREATE YOUR MICROFARM — 11

The 5 Main Steps of the Market Gardener Method — 12

Set Financial and Staffing Goals	14
Establish an Action Plan	16
Plan Your Farm Layout & Draw a Map	22
Prepare an Operations Calendar	24
Start Strong with the Right Equipment	26

APPLYING THE MARKET GARDENER METHOD TO A HOME GARDEN — 31

Grow Vegetables with the Market Gardener Method: 5 Steps	32
Understand and Analyze Your Garden Space	35
Determine Your Garden Layout	38
Prepare Your Crop Plan	42
Invest in Tools	44
Seeding	46
Transplanting	48
Monitoring and Care	50
Harvest Time	52

A YEAR OF VEGETABLES IN 12 MONTHS — 55

January	56
February	60
March	64
April	70
May	76
June	82
July	88
August	94
September	100
October	106
November	112
December	116

Acknowledgments	120
About New Society Publishers	120

Introduction: A Few Words About My Background

Drawing on principles from agroecology, permaculture, and entrepreneurship, I champion a modern form of nonmechanized farming, carried out on a human scale.

On a human scale means feeding many local families, while respecting the human and natural ecosystems in which we operate.

On a human scale means allowing market gardeners to make a decent living from their work, to run their businesses as they see fit, and to give themselves more time off than conventional farmers.

On a human scale means evolving through the use of technology but especially by relying on people and their skills and knowledge.

From Organic Farms...
I studied agroecology at McGill University's School of Environment in Montréal, where I met my wife and business partner, Maude-Hélène Desroches. At the time, we were both looking to create a new model for farming, one that would have a positive environmental impact. After graduation, we spent two years in New Mexico, USA, working on an organic farm and learning to be market gardeners.

Our microfarming aspirations were later fueled by a trip to Cuba where we spent time on *organopónicos*, fascinating urban farms that were established during the American embargo. During that era, after the fall of the USSR, the country developed a biointensive and urban agricultural model to ensure food security for the island's residents.

... to a Family-Run Microfarm

Back in Quebec in 2004, we acquired a small plot of 10 acres in Saint-Armand, in the scenic Eastern Townships. On this land, we experimented with our innovative approach to market gardening, which especially drew from the work of Eliot Coleman, an American market gardener who has been highly influential in the world of organic microfarming.

We built a 2-acre market garden, Les Jardins de la Grelinette, where we were able to test the first iterations of my method, now called the Market Gardener Method. It consists of crop rotation, the near-exclusive use of hand tools, organic growing practices, and shorter marketing channels, with direct sales made through CSA boxes and farmers' markets. At Les Jardins de la Grelinette, Maude-Hélène and I both worked full-time, and hired two farm workers (one full-time and the other part-time) to help with harvests.

Making 2 Acres Profitable

Success came quickly, both in terms of harvests and direct sales. After bringing in $33,000 in our first year, we earned twice that in the following year, and more than $110,000 in our third year of operation.

We were thus able to earn a living as market gardeners from almost the very beginning. Since then, our farm has continued to feed more than 200 families every year, offering roughly 40 types of vegetables, all grown on just 2 acres. Over the years, our harvests expanded and sales continued to increase. Eight years after starting the farm, I presented this farming model in a practical guide called *The Market Gardener* in 2014. The book was an instant success — over 250,000 copies have now been sold, and it has been translated into nine languages.

In 2015, with the support of a generous patron, I founded Ferme des Quatre-Temps in Hemmingford, Quebec, with the vision of creating a model for the future of ecological agriculture. On this 160-acre farm, we established a polyculture system in a closed-loop cycle, raising pasture-fed cattle, pigs, and hens, alongside a culinary laboratory. At the heart of the farm, 7.5 acres were dedicated to a market garden, where we applied the growing methods developed at Les Jardins de la Grelinette. It is here that I teach my apprentices the principles of productive and profitable market gardening.

The project was featured in a TV show called *Les fermiers*, which follows the evolution of Ferme des Quatre-Temps and its apprentices, who later start their own farms in front of the cameras. The show was a hit in Quebec and is now available on TV5 Monde and Apple TV.

In parallel, I worked to expand my methods to reach a broader, global audience. In 2018, we launched the Market Gardener Masterclass, a fully online course now available in over 90 countries. To further support this initiative, I founded the Market Gardener Institute with a clear mission: to educate the next generation of growers by equipping them with the knowledge, skills, and resources needed to become leaders in the organic farming movement.

The Institute has two key objectives: to teach best practices in market gardening techniques and growing methods, and to demonstrate that small-scale farming worldwide can not only be ecological but also productive and profitable. On a global scale, it's the number of farms, not their size, that holds the key to feeding the world.

Inspiring Change

My ambition is to drive meaningful change in society by promoting a way of farming that honors nature, supports communities, and empowers local farmers. I believe in a decentralized farming model, built farm by farm, as the foundation for a truly sustainable and resilient food system.

Since 2020, I have proudly served as an ambassador for the prestigious Rodale Institute, which researches regenerative organic farming practices in the United States and beyond. I am also honored to be the ambassador for Growers and Co., a company that develops tools and apparel for new organic growers. In 2023, we launched Espace Old Mill, a restaurant and market garden set in one location. The restaurant uses the best produce in the region, including harvests from our own farm.

What Is the Market Gardener Method?

While my approach may seem innovative, it is founded on practices that were first developed by 19th-century Parisian gardeners, who fed more than two million people through a network of thousands of market gardens—precursors to our modern-day microfarms—within the city of Paris.

These market gardeners applied remarkable ingenuity, skills, and knowledge to meet the increasing food demands of a city in the midst of urbanization and demographic expansion. They achieved this through organic, nonmechanized agriculture. From the mid-18th century to the 20th century, many books were written about the innovative practices of these market gardeners, whose technical feats were admired throughout Europe. But with the advent of modern practices, much of this know-how was relegated to the past.

As a result of mechanization, the advent of agronomic science, and improved refrigeration and transport that brought in fresh and inexpensive food grown abroad, farms grew in size, became less diversified, and took on a more technological focus—a trend that continues today.

Fortunately, these inspiring models led to the development of horticultural methods that have endured, and with the same objective: to grow sustainably, by maximizing vegetable yields without degrading soil quality. We now use the term "biointensive" to describe these methods. Unlike extensive agricultural operations, they continue to work on a human scale and offer farmers the opportunity to use little mechanization. Despite what some may believe, this approach is also profitable.

By working on only small plots of land, market gardeners can keep start-up investments to a minimum, compared to the funds needed for a conventional farm. Biointensive farmers also require a smaller workforce, doing the work themselves with the help of just a few employees. They also sell their produce directly to customers, avoiding commissions to intermediaries. These three factors allow market gardeners to start generating profits quickly.

Still, it's important to remember that working the land is never easy. While market gardeners can make a good living with this method, the first seasons are time-consuming and require a significant workload and financial investment. In this profession, nothing comes easy, and every dollar you earn is the fruit of your labor, the result of your organizational skills. That's why I always tell my apprentices to learn how to work smarter, not harder.

From a financial perspective, market gardeners should plan to start with an investment of $50,000 to $150,000, depending on whether certain assets are already available—such as a building that can be converted, access to abundant water, electricity, natural gas, or a vehicle. This amount does not include the cost of purchasing land, which can be amortized over 20 years, if needed. Renting is also an option that can prove very profitable, especially when the farm is located near a city or an affluent municipality, where land is expensive.

Regardless of experience and preparation, the first years of market gardening will be intense. Opening new ground, constructing greenhouses and tunnels, and setting up infrastructure (irrigation, washing and packing stations, nurseries, etc.) all take extra time and effort. However, once this phase is complete, market gardeners who have mastered their craft can do more than just make a living off a few acres—they can earn a very decent living.

This leads to another key principle I teach: your farm should work for you, not the other way around. Profitable and productive farming is possible, but you need to set it up for success.

Preface: Planning, the Key to Success

"A well-planned season is a successful season!" Every gardener should start their growing season with this maxim in mind. Being well organized and having a crop plan are both critical to a thriving vegetable garden. Each crop has its own specific needs in terms of timing, spacing, and care.

From early spring, you must plan what varieties to plant, when, and where in the garden. This step maximizes your available space and ensures a solid crop rotation plan, which is essential for disease prevention and long-term soil fertility. With an understanding of each plant's growth cycle, you can better time plantings for staggered and bountiful harvests throughout the season.

But planning is about more than just seeding dates. You also have to anticipate irrigation and nutrient requirements and prepare to manage pests and disease. With a solid garden plan, you can be proactive rather than reactive, implementing preventive strategies such as mulching, crop rotation, and selecting disease-resistant varieties.

In this guide, I offer a methodical approach to planning your growing season. You'll find tips for optimizing every plot in your garden, techniques for better anticipating and addressing the different needs of your crops, and strategies for extending your harvest season.

Your garden will become a productive and peaceful haven where each plant's needs are met, bringing balance to your growing space. Together, let's explore the secrets to a successful growing season. Wishing you a rewarding journey and abundant harvests!

Jean-Martin Fortier, market gardener in Saint-Armand, Quebec

Why Make a Plan?

Successfully growing about forty types of vegetables is no small feat. For professional and home gardeners alike, it's about taking an organized approach to crops, work, and time. Here's what good planning can help you do.

Manage Complexity
When it comes to vegetable microfarms, good planning takes many factors into account, such as the specific needs of each crop and changing weather conditions. This planning process helps growers to tackle these complexities with a streamlined approach and make good choices.

Reduce Mental Load
Market gardeners who don't plan will quickly become overwhelmed by the sheer number of tasks to be accomplished and this causes stress. Sound planning provides a long-term perspective that alleviates this mental load.

Optimize Time Management
Crop plans establish exactly when and where each crop will be planted. This information allows growers to optimize bed space and know in advance which tasks must be carried out for each crop. This results in workdays that are managed more effectively.

Navigate Seasons Decisively
Anticipating needs for supplies, materials, equipment, and labor allows you to set up a clear and coherent annual plan, make good choices and act effectively to run your microfarm with peace of mind for years to come.

Achieve Your Goals
Your objectives, whether they relate to finances, technical challenges, or staffing, can only be achieved through sound planning. This approach will improve the profitability of any farm. Even then, despite proper planning, you sometimes require flexibility, creativity, and common sense to address unexpected events and avoid being caught off guard!

Biointensive market gardening requires maximizing bed space and maintaining a consistent, well-organized planting schedule. In the field, this means using every bed in every plot at all times.

As soon as one vegetable crop reaches maturity, it is harvested, leaving the bed ready for the next planting.

Planning is a pillar of biointensive market gardening. Specifically, crop rotation is imperative to ensure a steady succession of harvests.

Market gardeners can maximize harvests and profits by optimizing space, carefully managing supplies, selecting appropriate vegetables, and planning well-timed crop successions.

By alternating between vegetable families and avoiding monoculture, you also reduce the likelihood of disease and pest damage, contributing to a more sustainable agriculture.

Using the Market Gardener Method to Create Your Microfarm

The 5 Main Steps of the Market Gardener Method

For professional market gardeners, I've made a plan for an entire year of vegetable production, which is also a source of inspiration for home gardeners!

The Steps

SET CLEAR GOALS
Define your financial and professional targets. Remember why you chose to grow vegetables and what your motivations are to continue doing so sustainably.

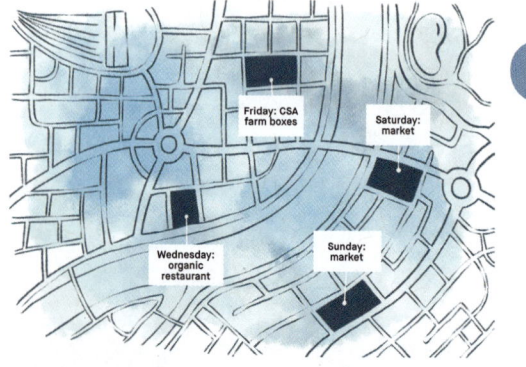

MAKE A PRELIMINARY PRODUCTION PLAN
Choose what crops to grow based on demand and profitability. Decide how much you want to grow and when each crop will be planted.

The 5 Main Steps of the Market Gardener Method

3 **MAP OUT YOUR GARDEN**
Draw an overview of your vegetable garden. Identify each plot, taking into account your needs and crop rotation plan. Maximize your space by integrating all infrastructure (tunnels, cold frames, tool shed, composting area, etc.).

4 **MAKE A CROP CALENDAR**
Create a calendar in which you schedule seeding, transplanting, and harvesting tasks. Consider timing requirements, days to maturity, and preparatory work (loosening and amending soil, etc.).

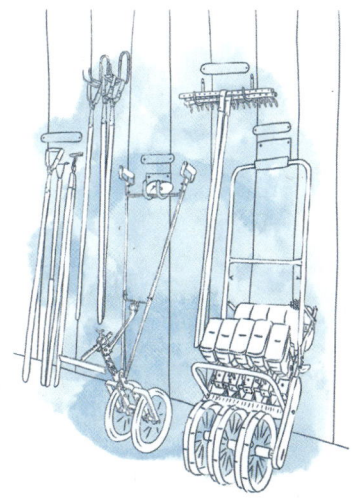

5 **ORDER SUPPLIES AND TOOLS**
Choose your seeds, materials, and equipment. Calculate the required amount then order a little more to provide a safety margin.

Tip from Jean-Martin Fortier
These steps are the foundation of crop planning and are critical for optimizing farm management practices, maximizing yields, and reaching your financial goals.

Set Financial and Staffing Goals

To establish a successful microfarm, every grower must start with a minimum financial investment and be surrounded by competent people.

Setting Clear Objectives

When crop planning, a critical first step is to establish sound financial and staffing goals. Start by developing a clear vision to determine the direction your microfarm will take in the coming season.

Your Financial Vision

Setting financial goals means creating a solid foundation for your market gardening season. Think of it as the treasure map guiding your journey. These financial targets help you answer a crucial question: where do you want your farm to take you over the coming season?

This is your opportunity to set clear, realistic financial benchmarks. Remember that it's important to be pragmatic and avoid wishful thinking.

By determining your yields, and thus your expected revenue, you will figure out whether your farm can meet your financial goals. This step is fundamental in assessing the financial viability of your operation and identifying any necessary changes.

Monitor sales trends so you are able to meet demand.

Staffing Requirements

Your financial goals are one piece of the puzzle, but your labor requirements are just as significant. Who will be a part of your team? How many people will you need to hire to hit your targets? What skills are required to help you successfully grow, care for, and harvest your crops?

Each member of your team must play a specific role in building a solid operation. By determining your staffing requirements, you'll know whether your financial goal is viable. If your team is already partially or fully established, this step helps you ensure everyone shares the same financial and staffing vision. This, in turn, will nurture better collaboration.

Know how to strengthen your team to address seasonal workloads.

Where Should I Start?

Start by determining how much revenue you must generate to achieve your goals. This is about so much more than dollar amounts—these numbers will serve as your compass. They answer crucial questions: How much income do you expect to generate for yourself and your team? What kind of profits do you want to make?

Once you have established your financial goals, assess whether they are realistic based on the size of your catchment area and your experience. Make sure your expectations are in line with your microfarm's environment.

Think about the workforce needed to achieve these goals. How many people will be required, and what skills should they have? Assess whether your current team has the skills and the capacity to handle your project. This might prompt you to pursue more education or training or consider hiring new staff.

Establish an Action Plan

In this step, you'll develop a strategic work plan for the coming season. This includes determining the types of vegetables to produce and how to market them and what equipment is required to grow them well.

Identify Sales Channels

To develop a strategy, you first need to identify potential points of sale. Consider your geographic location and whether there are local stores and restaurants that might appreciate fresh local produce. These can become excellent points of sale.

Farmers' markets offer a way to sell your products directly to consumers. Explore the markets operating nearby and choose ones that best suit your offerings. Explore whether veggie boxes exist in your area and whether you can join as a supplier or consider starting your own.

Establish an Action Plan

Local grocery stores and organic cooperatives, in search of fresh, local produce, may be open to partnering with market gardeners to supply vegetables. If your farm is located near a city or large town, consider opening your doors to customers for direct on-farm sales.

Decide What Vegetables to Grow

After deciding what vegetable crops to grow, you can determine the necessary bed space, staffing, and supplies.
Opt for high-value crops, those that deliver high returns per unit area. Include one or more uncommon vegetables in your offering to pique customer interest; displaying a few lesser-known items breaks the routine and makes you stand out from the competition.

Think about your hardiness zone, the length of your growing season, and any local microclimates that might affect crop outcomes.

Be careful not to overstretch yourself by offering a wide array of vegetables that you don't know how to grow well. It's better to do a good job growing a smaller variety of crops than to do a bad job growing lots.

Grow Crops Even in the Winter!

Remember to include vegetables that thrive in the fall and winter—brassicas, spinach, carrots, leeks, and turnips—and can be harvested even when temperatures drop below freezing. Lastly, invest in the right equipment for growing under tunnels or floating row covers to extend your shoulder season and harvest in the winter. This is critical for maintaining a steady stream of revenue in the off season.

Consider Storage Vegetables

Potatoes, beets, carrots, onions, garlic, and winter squashes are all sound choices. Plan ahead so you'll have enough to sell throughout the winter and make sure you have enough storage space.

To build customer's loyalty, offer freshly harvested vegetables such as carrots, leeks, or cabbage along with storage crops.

Some customers may wonder if your produce just came from refrigerated storage and therefore isn't quite farm fresh. To dispel these feelings and give your display a boost, consider growing extra-fresh vegetables such as mesclun, lettuces, or radishes.

Seeding Under Cover

Greenhouses and tunnels provide so many advantages for market gardeners. Although their initial cost may seem high, the structures quickly pay for themselves.

First, they allow growers to start crops earlier in the spring and keep them in the ground longer into the fall. This means you can produce vegetables that are less common in any given season and highly sought-after by consumers who will pay a premium price for them.

Second, extending your shoulder seasons with a greenhouse or tunnel allows many consecutive vegetable crops to grow in the same bed, thus increasing your total yields.

Finally, these structures protect crops from adverse and unexpected weather events like heavy rainfall, winds, and sudden drops in temperature.

As a result, you have more control over the vegetables' growing environment: you can manage the temperature, relative humidity, watering, airflow, and pests and diseases. In these closed environments, it's easier to stop the spread of fungal diseases or introduce beneficial insects. You will create optimal conditions for your crops.

Building a pond, moving soil, and installing buried infrastructure are tasks that require the occasional use of construction equipment.

Expect the Unexpected!

Planning also means anticipating unforeseen events and dealing with setbacks. To prepare, you should establish a budgetary roadmap that will guide you throughout the season. This document must account for spending on supplies—seedlings, seeds, and growing materials, as well as tools and all kinds of small equipment—and also provide for the acquisition and depreciation of more expensive equipment. Identify any anticipated major purchases such as new tools or permanent structures (greenhouses, tunnels, food-processing or packing rooms, storage facilities, etc.).

You should also plan for the cost to maintain and replace equipment that wears out over time like silage tarps and irrigation systems. Build these expenses into your budget so you'll have the funds when they are needed.

Additionally, bear in mind that you will likely have to deal with unforeseen events; a surprise frost, for instance, might require you to quickly buy protective equipment, repair damaged infrastructure, or replace destroyed plants.

Including this leeway in the budget will allow you to address the unexpected calmly and ensure that your operation is financially stable. By anticipating costs, you are better prepared to face all issues that might arise during the season. This, in turn, contributes to making your business more sustainable.

Accurately Assess Profitability

Hiring staff or investing in equipment can provide many benefits. It can allow you to increase your growing area, diversify your operations, and respond positively to new opportunities. For example, with a larger workforce, you could consider joining an additional weekly market or including more value-added products such as juices, sauces, or ready-made meals.

Similarly, acquiring new specialized equipment can streamline certain tasks and increase the efficiency of your operation. However, it's essential to carefully weigh the costs and benefits of these investments to ensure they contribute to making your business more profitable. A sound financial plan and thorough analysis can help you make the right decisions about managing operations while maintaining satisfactory and sustainable profitability.

Tip from Jean-Martin Fortier

Everything that happens in your garden is determined by the operations calendar you create during early-season planning. This serves as a roadmap guiding you towards success and profitability, so take the necessary time to carefully craft it and make it your trusted partner for the coming season.

Plan Your Farm Layout & Draw a Map

On paper, this map allows you to visualize the layout of the farm and organize the distribution of plots around existing buildings.

The results of a soil analysis help you better understand its composition.

Once you've chosen what crops to grow and their sales channels, it's time to draw a map of your farm layout and organize the different areas: vegetable plots, washing and packing stations, aisles and pathways. Putting these details on paper makes it easier to plan and organize the work.

This overview is essential for effectively managing the complexities of crop successions, determining locations for vegetables, establishing planting and harvest schedules, and optimizing the use of infrastructures like tunnels and greenhouses. The main objective is to optimize use of available space while providing ideal growing conditions for each crop.

Before drawing up your overall plan, it's a good idea to conduct a soil analysis.[1] This will help you understand its characteristics and select crops accordingly. Bountiful harvests are a sure bet when you have a balanced living soil.

When creating your map, you have to consider multiple factors that may be tricky to bring together. First, group vegetables according to their needs (irrigation, shading, and plant protection), which will make crop maintenance much easier. For example, if you're planning to set up a sprinkler system, avoid planting crops nearby that are prone to disease when the foliage gets wet such as summer squashes. To spend less time weeding, put direct seeded crops in clean beds that were previously covered with a silage tarp (occultation tarp). To limit disease and insect pressure, implement a crop rotation plan so that in any given area you don't plant the same veggie or two varieties from the same family.

Second, include cover crops like clover or mustard, in crop planning. They help to improve soil quality by fixing nutrients while reducing weed pressure.

Always Evolving

As beautiful as your map may be, it's not a work of art! It's a real-life working tool that must be updated as your operation evolves. Each new season requires adjustments to plots, beds, and equipment. If you regularly update it, the map will remain a valuable visual guide for proper management of your microfarm.

---

[1] See *Living Soil* in the Grower's Guides from the Market Gardener collection.

Prepare an Operations Calendar

This document improves outcomes for any farm because it outlines the tasks to be performed for each crop and organizes the work for the entire team.

Having drawn up the layout of your plots on a map and divided them into beds, you can create an operations calendar. Throughout the season, this roadmap for your crops will be useful for improving efficiency in sowing, planting, and crop maintenance.

Start by writing down the intended planting dates for each crop, taking local climate into account. Use weather forecasts and average temperature data to better adjust these dates. Next define anticipated harvest dates for each crop based on the variety's days to maturity. Make sure you allow enough time for harvesting, preparing, and packing vegetables before they go to market.

You should also integrate all crop and soil maintenance tasks, including weeding, irrigation, and phytosanitary treatments. Allow yourself some flexibility in case of unforeseen events such as bad weather or insect pest infestation.

Your operations calendar is an adaptable tool that you can tweak as the season progresses. It helps you anticipate labor, supplies, and equipment needs for every stage of each crop's growth. Ultimately, a carefully crafted calendar will significantly contribute to a successful gardening season.

An operations calendar that is kept up to date allows you to stay on track from seeding to harvest.

Staying on Track

After finalizing your operations calendar, you must stick to it as much as possible to maintain an efficient and consistent production all year long. Update it by recording all operations undertaken, from planting and harvest dates to irrigation frequency, fertilizer inputs, and phytosanitary treatments. This practice allows you to monitor each crop's development.

If you're working with a team, make sure everyone knows the schedule as it outlines the operations to be done for each crop and specifies the tasks to be completed for the week, or even for a given day.

A well-managed calendar will support your farm's overall success by ensuring effective management of time, resources, and crops throughout the growing season.

Tip from Jean-Martin Fortier

Make sure you take notes throughout the season and record any lessons learned in your operations calendar.
It's completely reasonable to make mistakes when starting up a microfarm. What's not acceptable is repeating them year after year!

Start Strong with the Right Equipment

The final step in setting up a microfarm involves purchasing equipment and supplies.

To get off to a strong start and not waste time, make sure you aren't missing anything. Planning ahead simplifies the ordering process. With a clear understanding of your entire operation and growing plan, you can anticipate the necessary equipment and supplies, ensuring you have them all before the season begins, which avoids having to rush out to buy a tool or materials, especially when extensive fieldwork is underway.

In the previous steps, you listed what was needed and how much: tools, seeds, potting mix, containers, floating row covers, etc. All that's left is to decide which varieties, brands, or models are the best for you and put in an order.

Keep in mind when ordering that specialized equipment for small-scale vegetable farming may involve particularly long delivery times.

Selecting Seeds

Seeds play a central role in farm operations, and for market gardeners, selecting the right seeds is a crucial step that determines crop quality, disease resistance, harvest yields, and adaptation to local growing conditions. We therefore recommend carefully consulting catalogs from different seed producers and comparing the advantages and specific characteristics of their varieties.

Tools

Which tools are right for you? This depends on your types of crops, size of operation, and intended growing methods. From soil preparation to harvesting, every step requires specific tools to help you work efficiently and effectively. When deciding on a particular purchase, always ask yourself the following question: can this tool save me enough time to make it a worthwhile investment?

Before getting a greens harvester, you should determine exactly what your needs are for this piece of equipment.

While certain hand tools like broadforks, rakes, hoes, tilthers, and seeders are essential for smooth operations, you might legitimately ask whether you truly need a walk-behind tractor or greens harvester. On our farm, the mesclun harvester significantly decreased time spent harvesting and increased the profit this crop generated on a weekly basis, both of which more than justified the purchase.

Sharing certain equipment such as a walk-behind tractor can be a way to reduce investment costs.

It's best to keep small supplies on hand for when the need arises: seeds and bags for storing them, twist ties and rubber bands for bunching veggies, paper bags or containers for packaging fragile vegetables, etc.

Consider harvest bags, as they provide flexibility and are easy to use.

Don't Forget Growing Materials and Supplies

Fertilizers, potting soil, row covers, stakes, twist ties, and harvest bags and crates are some of the many accessories that complement your garden tools. These products, which regularly have to be replaced or maintained, must meet the needs of each crop. To market your produce, you also will require packaging, such as trays and paper bags, as well as rubber bands to bundle vegetables. Don't disregard these small supplies as they help showcase your produce display. Your choice of packaging and its source also reflects your farm's environmental stance.

A Final Word of Advice

It's a good idea to keep an inventory of all materials and supplies in order to know exactly what you have on hand, how much was used, and when to reorder stock.

Tip from Jean-Martin Fortier

Seeds intended for professional market gardeners are not always identical to those sold to home gardeners. While it's essential for all growers to select varieties that are suited to their soil and local climate, with proven yields and disease resistance, professionals must also consider profitability and harvest uniformity to ensure customer satisfaction. We therefore recommend modern varieties, often labeled F1, rather than heirloom varieties.

Though the biointensive Market Gardener Method for growing vegetables was originally developed for professional market gardeners, it has proven to be a real source of inspiration for home gardeners.

To transform your home garden into a fertile haven producing bountiful harvests, you just need to understand the fundamental principles, scale them to your garden size, and then put them into practice. When you work in sync with nature, treat the soil with respect, practice succession planting, and select varieties that are suited to each season, your garden will yield fresh healthy vegetables nearly all year-round.

This is what the biointensive Market Gardener Method is all about—it is so much more than just a way to grow vegetables. It's a philosophy, a worldview that all home gardeners can adopt who want to grow their own vegetables to better understand what they are eating, take an active role in protecting the planet, and reap the fruits of their labor.

Applying the Market Gardener Method to a Home Garden

Grow Vegetables with the Market Gardener Method: 5 Steps

A vegetable garden is more than just beds and plants; it is an ecosystem that combines nature, science, and passion to grow veggies. Microfarming practices apply even on the scale of any size home garden.

1. SET GOALS
The purpose of a home garden is to feed a family or household, so the amount of food you grow should be tailored to the number of people to be fed. This will then determine the required garden area, depending on whether you grow storage vegetables such as potatoes and cabbages that take more bed space.

2. OPTIMIZE SPACE
Since home gardeners typically have limited space, they must streamline their layout of aisles and pathways, technical spaces (sheds, compost), and growing infrastructure (tunnels and frames) to maximize available square footage for garden beds. Don't forget to include herbs, berries, and fruit trees in your plan.

Seedlings, vegetables ready for harvesting, a low tunnel, berry bushes, and fruit trees coexist in a home garden.

Grow Vegetables with the Market Gardener Method: 5 Steps

3 MAP OUT YOUR GARDEN
Draw each area of your garden on a sheet of paper to get an overview of the space and how the different parts (beds, technical areas, shelters) fit together.

When updated annually, this map makes it easier to manage the tasks for each bed, from seeding and transplanting to crop rotation. It's scalable, and depending on the number of household members, you can make areas bigger or smaller.

4 PREPARE AN OPERATIONS CALENDAR
After planning your schedule for seeding, planting, and crop maintenance tasks (including scouting for pests and disease), you should write it into an annual operations calendar. As a daily aid, this calendar will be immensely helpful. More importantly, it allows you to assess whether, over the year, the time devoted to the garden is compatible with your other obligations (family, work, leisure). You can develop an operations calendar in a paper planner, downloadable spreadsheet, or use specialized websites and software.

5 MANAGE YOUR BUDGET
Managing a home garden and growing vegetables require some financial investments: seeds and plants, materials, tools, and other equipment that are easily purchased from specialized stores, by mail order, or online.

It's essential to calculate these expenses and include them in your annual garden budget. The aim here is to ensure that your home production is costing less overall than buying the same vegetables at a market or store.

Grow Slow

It's easy to dream big and get carried away with enthusiasm about starting a garden. Remember that when you apply the Market Gardener Method, a well-managed vegetable garden—however small—can be incredibly productive. So start small with a limited, manageable area to avoid feeling overwhelmed by the care and harvesting that is required.

Take time to identify your household's needs, quantity of vegetables to eat fresh, and how much to preserve. Your home garden must evolve gradually and realistically—that is the key to success. Over the years, as needs change and you gain experience, you can always increase the garden size, diversify crops, and expand the range of varieties. Vegetable gardening is a long-term journey, so make sure you enjoy every step.

Understand and Analyze Your Garden Space

Sunlight, soil type, climate, and weather events are essential factors to consider before starting a vegetable garden.

Sunlight

The amount of light that reaches each plot in your garden is fundamental to growing vegetables. Watch where the sun shines throughout the day in every season and then choose locations for different vegetables based on sunlight. Some plants prefer full sun, while others thrive in partial shade.

You should also identify areas that are shaded by trees, buildings, and other nearby features. Note these shady zones will change significantly depending on the time of day and month.

Wind Effect

Wind can be a determining factor in your vegetable garden design as strong winds can quickly dry out soil, damage delicate crops, and disrupt plant growth. To mitigate these impacts, it's a good idea to create sheltered areas to protect crops. You can plant hedgerows or install fences to act as windbreaks.

You can plant hedgerows or bushes around your vegetable garden to limit damage from wind and cool breezes.

Hedgerows are an excellent option because they also provide shelter for beneficial wildlife.

Installing multiple water spigots throughout the garden eliminates using excessively long hoses and makes crop irrigation easier.

Choose trees and shrubs suitable for your climate and local growing conditions, making sure to plant them far enough from your beds to avoid creating too much shade in later years.

Assess Irrigation Needs

Water is a vital resource for vegetable growers. Assess each crop's irrigation requirements and identify alternative sources to your drinking water such as rainwater collection, a well, or a pond. You can go green and save money by installing a rainwater collection system to collect and store run-off for later use. Place rainwater collection tanks below any downspouts or drainage features on existing buildings. This allows you to irrigate crops without using potable water from your municipal network, thereby conserving local water resources.

Installing one or more spigots in the garden provides easy access to water, which is essential for irrigating crops, especially during droughts. Make sure to strategically locate these access points near your beds to allow for shorter hoses and save endless trips with a watering can.

Microclimates

Microclimates, whether warm or cool, must be taken into account when planning a vegetable garden. Consider your local weather characteristics—winds, late spring frosts, and early fall frosts—and adjust the orientation of beds or location of tunnels and cold frames accordingly.

Understanding and Improving Your Soil

Soil is the foundation of all vegetable gardens. It varies from one location to the next, so you and your neighbor may not have the same type of soil. Because it has a major impact on plant growth, you have to know your soil's composition really well.

We recommend starting with a soil analysis, by either doing easy at-home experiments or sending a sample to a testing lab. The analysis will determine your soil's texture, pH, and fertility. This crucial information helps you select crops that will grow best in your garden. It also allows you to improve the soil's characteristics and fertility by amending it accordingly with compost, manure, sand, or other additives.

If you're new to soil analyses, seek assistance from an expert who can provide invaluable advice on interpreting results and implementing a soil improvement plan. By developing a better understanding of your soil and taking steps to improve it, you will create optimal conditions for your vegetable garden to thrive.

One major source of organic matter is composted green waste such as grass clippings, dead leaves, and small branches. This composted material is rich in organic matter and nutrients. It improves soil structure and texture, enhances water retention capacity, and fosters microbial activity. This quickly creates a beneficial cycle, resulting in healthier and stronger vegetables as well as higher yields. For instance, by incorporating a layer of compost while preparing beds, you create ideal conditions for good plant growth and contribute to the garden's long-term fertility.

For more information on all aspects of soil, see *Living Soil: A Grower's Guide,* in this book series.

Adding compost every year will improve soil life.

Tip from Jean-Martin Fortier

Initial observations are crucial when planning a home vegetable garden. If your site doesn't meet all criteria, assess what changes you can reasonably and feasibly make. For example, you might consider pruning or even falling large trees that shade the garden. Of course, this is only possible if they are on your land. If they are next door, you must get the neighbor's consent, which is not a given! In such cases, you could consider moving your garden or adjusting the size.

Determine Your Garden Layout

Having mapped out the different areas in your garden, your logical next step is deciding the location of beds, aisles, plots, shelters such as tunnels or greenhouses, and composting areas.

A Well-Structured Vegetable Garden

It's a good idea to set up plots and beds with identical dimensions throughout your vegetable garden. This uniformity provides many advantages for efficiently managing the space and equipment. When all plots are similar in size, you can use the same equipment—row covers, mobile tunnels, drip irrigation systems, etc.—everywhere without needing to adjust them every time.

It's also much simpler to implement crop rotation when you know exactly how much space is available for each type of crop from one season to the next.

Aisles & Pathways

Well-defined pathways (along plots) and aisles (between beds) make it easier to get around the garden to maintain, monitor, and harvest crops. Pathways should be wide enough, at least 24 to 36 inches (60–90 cm), to accommodate a wheelbarrow. Aisles between beds can be narrower, roughly 12 to 16 inches (30–40 cm) wide.

Narrow aisles between beds allow growers to access crops. They connect to wider pathways that make it easier to get around, especially with wheeled implements.

Compost

Setting up a composting area strategically near your vegetable garden is an essential step that allows you to transform plant waste into fertile, nutrient-dense compost, a major asset for feeding the soil. This location minimizes transportation when amending beds with compost.

If you'd rather not have an unsightly traditional compost heap at the back of the garden, consider surface composting. With this more discreet method, decomposing organic matter is spread directly onto the soil surface to gradually release nutrients while maintaining the aesthetic appeal of your garden space.

Crop Shelters

A greenhouse or tunnel can significantly improve the productivity of your vegetable garden by extending the growing season and shielding crops from bad weather, pests, and disease.

However, you may want to get two or three years of experience before making such a costly investment. Conditions can change from year to year, and this time spent learning and reflecting is an opportunity to better assess your needs.

Set aside a dedicated composting area, so you can recycle green waste.

Determine Your Garden Layout

In a small greenhouse, about 85 to 110 square feet (8–10 m²), you can produce seedlings, grow early crops, and provide sheltered space for delicate vegetables.

Make sure your building materials are labeled and certified as environmentally friendly and safe for plants, animals, and gardeners.

By planning and laying out your garden wisely, you can create an environment that supports healthy plant growth. You'll also limit potential issues arising and increase your odds of success by making good use of available space and resources.

Tip from Jean-Martin Fortier

Before finalizing your garden layout, mark the locations of the beds and push a wheelbarrow around them to ensure the spaces are functional. This highly pragmatic approach allows you to make adjustments according to the terrain. You're better off investing time early on rather than laboriously fixing issues later.

Prepare Your Crop Plan

While professional market gardeners focus on profits when planning, home gardeners can be more flexible, allowing for whimsy and choosing rarer vegetables or less productive varieties.

LIST YOUR FAVORITE VARIETIES
Start by listing the vegetables and herbs you want to grow in your garden. Think about your personal preferences and local growing conditions.

Research your hardiness zone to find out which crops are best suited for your climate as this will determine planting dates and harvesting times.

MAKE A WEEK-BY-WEEK CALENDAR

Create a calendar—on paper or using software— to indicate dates for seeding, potting up, transplanting, crop maintenance, and harvesting for each crop each week. Use color codes or symbols to improve readability.

Some vegetables have specific needs in terms of watering, fertilization, cold protection, and pest prevention, so note them in your schedule.

BE FLEXIBLE

Pay attention to the weather and any other less predictable factors that may require changing your schedule. You might have to shift dates for certain field operations as the season progresses. Afterwards transfer these changes to the calendar, which will serve as a reference for future years.

Invest in Tools

To run a successful vegetable garden, it is essential that you have the right equipment.[1] It allows you to work efficiently, care for crops with precision, and minimize unnecessary efforts. Good tools make for good gardeners!

BROADFORK
Loosens the soil without turning it over or harming the microorganisms living underground. The broadfork is central to the Market Gardener Method.

RAKE
Used to move debris and clumps of soil after loosening a bed or to level and smooth the surface before seeding or transplanting.

---

1 See *Vegetable Garden Tools* in the Grower's Guides from the Market Gardener collection.

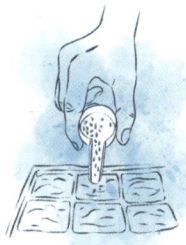

DIAL SEED SOWER
Allows for high-precision sowing as growers can adjust the rate of seed distribution. It's a great tool for seeding into open flats and plug flats.

DIBBER
Used to dig holes of all sizes, transplant seedlings, fill holes, or move small amounts of soil.

WATERING CAN
Makes it easier to water at the base of a vegetable, keeping the foliage dry. The sprinkler head (rose) provides a gentle spray, especially for seedlings and seedbeds.

HOE
Loosen the soil surface, kill weeds, and hill young plants.

PRUNING SHEARS
Used to prune and care for your plants, making clean and precise cuts.

ROW COVER
A white piece of fabric used to shield crops from significant temperature fluctuations. Insect netting also protects plants from harmful insects while allowing air and light to pass through.

HARVEST KNIFE
Used to cut stems when pruning or harvesting vegetables.

Tip from Jean-Martin Fortier

Don't underestimate the importance of having the right tools even if they're expensive; they will save time and reduce fatigue. Make sure your tools are well maintained and sharpened so they'll remain effective for a long time. And don't forget protective gear such as gardening gloves or kneepads that can cushion your joints when working in the soil.

Seeding

For gardeners, starting crops from seed is both gratifying and highly cost-effective.

Seeding into plug flats.

Seeding Under Cover

Seeding under cover means sowing crops (usually delicate or early vegetables) under the shelter of a glass or clear plastic structure such as a greenhouse, mini greenhouse, or a cloche that provides a warm, bright environment and keeps the cold air out.

Crops are sown into containers such as open flats, plug flats, or pots filled with a blend of compost and sand designed for seed starting. If you purchase this potting mix from garden centers, make sure it is clean and free of seeds, weed fragments, and disease. You can make your own by blending regular potting mix with sand in equal parts.

Seeded trays and pots kept under cover with the right temperature, humidity, and light will germinate in a few days or weeks, depending on the crop. Once seedlings have a few leaves, they are ready to be potted up or transplanted into the garden after a hardening off period.

Potting Up

A few weeks after sowing a crop, transplant seedlings into individual pots to promote root development and prepare them for transplanting. This entails gently removing a seedling from its tray or pot, keeping the root ball as intact as possible, and replanting it in a container filled with potting mix, being careful not to damage the roots. Then tamp the soil, water the seedling, put the container in a bright sheltered space for 3 to 4 weeks.

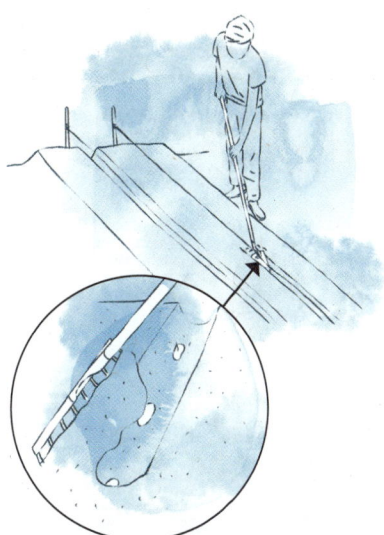

Direct Seeding in Your Veggie Garden

Direct seeding refers to the practice of sowing a crop directly in the ground, dropping seeds into a hole or furrow. This applies to vegetables that don't transplant well such as certain root vegetables and greens. Crops that are direct sown must be well suited to this process and your local climate. If the weather is too cold or there is a risk of frost, protect the crop with a floating row cover, low tunnel, or cloche.

Tip from Jean-Martin Fortier

Whenever possible, I always opt to sow as many seedlings as I can under cover. This provides many advantages, most notably accelerating plant growth. Seedlings grown under cover become strong plants that can handle being transplanted, which speeds up production.

Transplanting

After sowing, this step consists of planting seedlings in the ground where they continue to grow until they are harvested.

Water seedlings just before transplanting them.

Carefully preparing beds makes planting easier.

Preparing to Transplant Seedlings

Make sure to prepare your soil properly. It must be deeply loosened, weed-free, amended with compost or organic fertilizer, and then carefully leveled on the surface. To promote better growth, water seedlings before transplanting them.

Transplanting

When planting, follow the recommended spacing between plants and rows. While dense planting may save some space, it can make hoeing the bed more difficult. Proper spacing ensures that each plant has enough space to grow without excessive competition

After transplanting seedlings, install the necessary equipment such as stakes to support tomatoes.

Transplanting

In the field, dig a trench (when planting potatoes, for example) or a hole that will fit the root ball you intend to transplant. Plant the seedling, being careful not to damage the roots, fill in any gaps, then tamp the soil.

Make sure you follow the recommended planting depth for each crop as planting too deep or too shallow can hinder growth.

Water thoroughly and, as a precaution or if necessary, set up insect netting or a floating row cover to protect the crop from pests or cold weather.

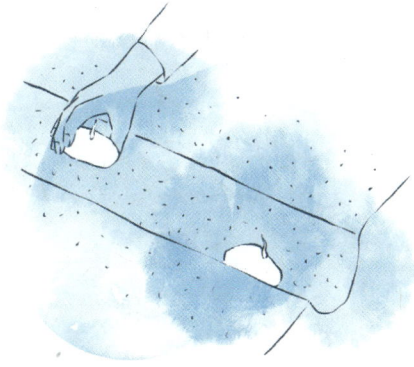

The type of hole, trench, or furrow depends on the vegetable. Here potato tubers are planted in a trench.

Covering the soil with a straw mulch after planting helps keep the soil moist and reduces weed pressure.

Tip from Jean-Martin Fortier

Mulching the soil right after planting is now common practice as it improves growing conditions for vegetable crops, keeping the soil cool, saving on watering, and limiting weed growth. This involves spreading a layer of straw, dead leaves, or grass clippings to cover the soil surface around your plants. Plus, a plant-based mulch gradually supplies nutrients to the soil as it decomposes.

Monitoring and Care

After putting a crop in the ground, it's time to begin monitoring and maintenance, which includes daily care such as watering and less frequent operations like weeding.

Covering crops with insect netting protects them from insect pests and prevents birds from eating tender shoots.

Pest Management

As a preventive measure and to avoid spraying your crops, cover plants with insect netting to keep out pests such as aphids and various flies.

Working with Beneficial Insects

Nurture biodiversity by planting hedgerows near the vegetable garden and leaving certain areas of the garden untended or mostly hands-off that will host beneficial pollinators such as ladybugs, hover flies, and bees.

Weeding

Hoe beds regularly to keep the soil clean, killing weeds before they have time to spread and preventing them from competing for nutrients with vegetable crops.

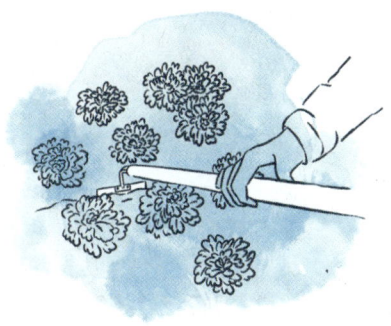

Irrigating

Be strategic when irrigating to avoid wasting water and issues related to overwatering, which suffocates roots. Direct water towards the base of each plant and try not to wet the foliage. In the summer, irrigate either early in the morning or in the evening to minimize water loss to evaporation.

Fertilizing

Amend soil with compost or thoroughly decomposed manure to provide the nutrients that plants need to grow and develop fruits. Choose organic fertilizers, either animal-based (manure, horn meal, etc.) or plant-based (nettle tea, etc.), that nurture microorganic soil life.

Harvest Time

Vegetable harvests are the culmination of many hours spent monitoring and maintaining crops daily. For gardeners, they are rewarding, justifying all their work that went into growing the vegetables.

Harvesting at the Right Time
Each vegetable has an ideal harvest window for getting the best taste and quality. To decide the right time to harvest, assess its size, skin color, and softness.

With some crops, a few days pass between harvests, while others must be picked more regularly. The latter is true for vegetables that provide continuous yields such as tomatoes, summer squashes, and beans. More frequent picking promotes new growth and extends the harvest.

Using the Right Tools
Make sure you always have the right tools on hand for harvesting vegetables such as a knife, scissors, or pruning shears, all with clean sharp blades. To store and transport your produce, use baskets, wooden crates, plastic trays or bins, or canvas bags.

Avoiding Waste
Harvest surpluses are common in home gardens. To avoid this problem, consider harvesting vegetables when they are still young and small—and often tastier. It's better to pick mini eggplants or thin early beans rather than large stringy fruits.

If you have a harvest surplus, consider canning, dehydrating, or freezing produce. This does require more time and equipment, but the benefits will quickly outweigh the investment.

Tip from Jean-Martin Fortier
One of the biggest lessons I've learned in my years as a market gardener is that it's important to grow slightly larger quantities in order to preserve some of the harvest to enjoy in the offseason. While this approach won't result in total food self-sufficiency, it's a significant contribution. Therefore you should choose vegetables suitable for canning and have the equipment, space, and skills needed to process them.

The Market Gardener Method of biointensive gardening emphasizes the importance of careful crop planning and consistent succession planting over the months of the season. While the method was designed for professional market gardeners, it also serve as inspiration for home gardeners organizing seeding, planting, crop care, and harvesting.

But first, you must understand the basics and how to implement the method. From soil preparation and the first seedings, to planting, weeding, pruning, and harvesting, the workload in every garden follows the rhythm of the seasons. Some periods, especially spring and early summer, are particularly busy and can discourage gardeners who might feel overwhelmed by the tasks to be addressed and harvests that can't wait! As the months go by, your perpetual calendar is a valuable guide, indicating what needs to be done and helping you to anticipate upcoming operations. This system, inspired by the Market Gardener Method, goes beyond simply tracking dates. It embodies a philosophy, an invitation to live in harmony with nature and to transform your veggie garden into an inexhaustible source of healthy and delicious crops. The calendar that follows is geared to more temperate climates, and while the tasks and cycles remain consistent, you can adapt the calendar to your specific climate.

A Year of Vegetables in 12 Months

January

"In January, vegetable gardens often rest under a soft blanket of snow, but dreams of bountiful harvests drive growers to begin sowing their first crops. Every passing day brings us a little closer to spring and a time of abundance that we prepare for in the first month of the year."
—Jean-Martin Fortier

Seeding Under Cover

· **Sow first lettuces.** Using plug flats filled with quality potting mix, sow lettuce seeds then keep trays under cover to help seedlings grow strong. In January, watch out for a lack of light and high temperatures that can make plants leggy.

- **Pre-sprout new potatoes.** For planting from mid- to late-February, start sprouting potatoes indoors by putting the tubers in a cool, bright room.

Planting Under Cover

In January, there's limited planting in vegetable gardens due to winter conditions, especially in regions with intense cold.

- **Transplant bok choy seedlings.** On sunny days, transplant bok choy seedlings in spaces vacated by previous harvests, covering them with a tunnel or floating row cover to create a favorable microclimate.

Don't Forget

- **Monitor stored vegetables.** Regularly inspect storage vegetables (potato, celery root, beet) and remove any damaged ones. They can quickly rot and contaminate everything else.

- **Tackle crop planning.** Use this time to draw up a detailed garden plan. Organize the layout, indicating the location of each upcoming crop. Consider crop rotations to prevent soil depletion and the spread of disease.

- **Provide frost protection.** If your region experiences heavy frosts, make sure to properly protect winter crops with cloches, low tunnels, or row covers to avoid frost damage. With leeks and brassicas, a layer of dead leaves or straw mulch will suffice. Remember to check local weather forecasts and adapt your strategies accordingly to avoid any exposure to frost.

An overview map of your vegetable garden can help you plan crop rotations.

- **Prepare soil.** Weather permitting, you can start preparing plots by running a broadfork down the beds and adding compost and other amendments to meet your crops' needs for the growing season.

- **Build support systems.** If you plan to grow climbing vegetables like peas or beans, prepare support structures (steel trellises, stakes, etc.) to train voluble plants and support fruits.

- **Maintain tools.** Use this quieter period to inspect and care for your gardening tools. Clean them, sharpen any blades and metal components, and replace worn parts.

- **Take stock of inventory.** Compile an inventory of seeds to identify which varieties you have in stock. Check expiration dates. Place an order if you notice gaps in seed supply or want to try new varieties. Make sure to order early enough to ensure they will arrive in time for spring sowing dates.

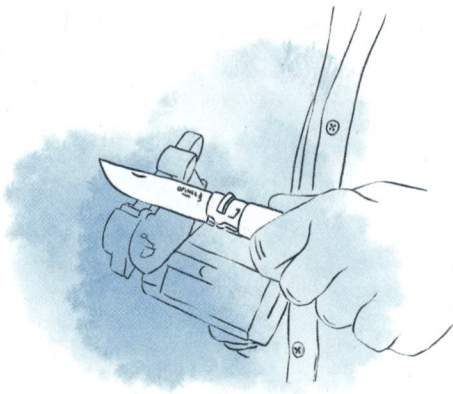

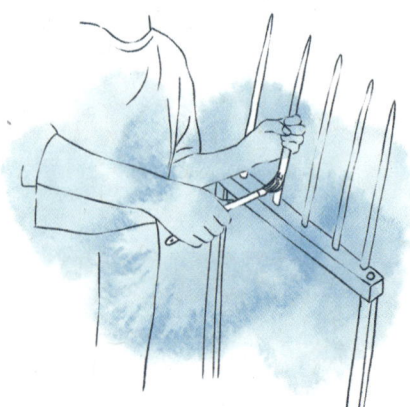

- **Order equipment.** If you need gardening equipment or other supplies, this is the perfect time to place orders so they'll be available when the gardening season begins.

- **Assess previous successes and failures.** Review your experiences in the last gardening season. Identify what worked well and what could be improved. This exercise will help you progress as a gardener.

- **Get informed.** Use this winter time to read gardening books and catalogs. You can also take online courses and watch videos to improve your skills.

Harvests

At this time, it is possible to harvest winter vegetables that survived the cold including:

- Spinach
- Mâche (corn salad) (above)
- Chicory (escarole and curly endive)
- Kale
- Leeks
- Crosnes (Chinese artichoke), dug up on demand
- Salsify (vegetable oyster) and scorzonera (opposite)
- Jerusalem artichokes
- Asian greens
- Mesclun

Tip from Jean-Martin Fortier

I can't overemphasize the importance of meticulous planning for the coming seasons. In January you should take the time to sketch out a map of your vegetable garden, choose what crops to grow, select vegetable varieties, and order seeds. Additionally, make sure you have all the necessary tools and equipment ready.

February

> "During this second month of the year, we are more active and the pace picks up with continued sowing of summer crops under cover: sweet peppers, hot peppers, eggplants. In the garden, nature is slowly waking up and the list of tasks is gradually increasing."
> —Jean-Martin Fortier

Seeding Indoors

· **Sow sweet peppers, hot peppers, and eggplants.** Fill open flats or plug flats with good-quality seed starting mix. Leave seeded containers on a windowsill near a radiator in a bright room. Ideal germination temperatures range from 68 °F to 75 °F (20–24 °C).

· **Sow celery.** In open flats filled with potting mix, sow celery seeds at 75°F (24°C) and place trays in a bright location. Keep the soil moist until the first shoots appear.

- **Sow cabbages and broccoli.** Fill a plug flat with potting mix, drop 1 seed per cell at ¾ inch (2 cm) deep, fill the hole, lightly tamp the soil, and water regularly until germination.

- **Sow lettuces.** At this stage, remember that you will be transplanting them later. Sow seeds into open flats, lightly tamp the soil, and keep it moist until germination. When the seedlings have 3 or 4 healthy true leaves, pot them up into plug flats.

- **Sow artichokes.** Fill pots with light, well-drained potting mix and gently push seeds into the soil. Leave the pots in a warm, bright room until seedlings are 6 to 8 inches (15–20 cm) tall and ready to be planted in the garden.

- **Sow kohlrabi and fennel.** In plug flats, sow kohlrabi and fennel seeds, tamp the soil, and water. Transplant them into a low tunnel, high tunnel, or cold frame once the risk of frost has passed.

- **Prepare sweet potato slips.** Lay tuber segments side by side on flats filled with potting mix then place them in a bright space kept between 72°F and 77°F (22–25°C). In this germination phase, spray the slips (sprouting sweet potato) with warm water to keep the soil moist and to stop the tubers from drying out.

Seeding Indoors

- **Sow baby carrots.** In well-drained, loose soil, with a well-prepared fine seedbed, sow carrots in rows, cover with fine soil, and tamp with the back of a rake. Water regularly to keep the soil moist until germination.

- **Sow early radishes.** Make shallow furrows 6 to 8 inches (15–20 cm) apart, drop seeds into the bottom every 1 to 2 inches (2–5 cm), cover lightly with soil, water, and harvest about 3 to 4 weeks after germination.

- **Sow spinach and mâche (corn salad).** Sow every 2 inches (5–6 cm) into ¼- to ½-inch (1 cm) shallow furrows 8 inches (20 cm) apart. Fill in the furrow and tamp with the back of a rake to ensure good soil contact with the seeds then water.

- **Sow turnips.** Loosen the soil and make furrows 12 inches (30 cm) apart. Drop seeds into the bottom about 1 inch (2–3 cm) apart. Water as soon as the soil surface begins to dry and until germination.

- **Plant new potatoes.** In loose, well-drained soil, plant sprouted tubers every 8 to 12 inches (20–30 cm) in shallow furrows 12 to 14 inches (30–35 cm) apart.

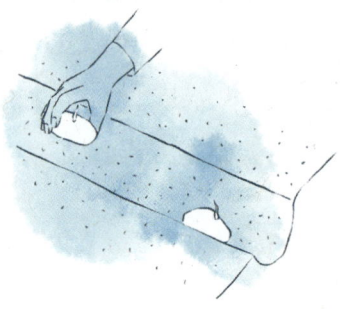

- **Sow Brussels sprouts.** Prepare the soil, sow into furrows 8 inches (20 cm) apart, and lightly cover with fine soil. Water regularly, but don't overwater, to keep the soil cool and moist. Use a floating row cover to protect seedlings from frost and, when plants have 4 to 6 leaves, pull them out and transplant into rows 16 to 20 inches (40–50 cm) apart.

- **Leeks.** Sow in rows 6 to 8 inches (15–20 cm) apart and cover lightly with fine soil. Once the leek shafts are as thick as a pencil, carefully pull them out and replant every 6 to 8 inches (15–20 cm) in rows 12 inches (30 cm) apart.

- **Parsley.** In pots or flats filled with a blend of potting mix and sand, sow seeds at a shallow depth, water lightly, and place in a greenhouse or a bright location on a sunroom until the plants are 2 to 3 inches (5–8 cm) tall. Be patient as parsley can be slow to germinate. Transplant seedlings into your garden when the risk of frost has passed.

Direct Seeding

- **Sow beans and peas.** Sow 3 rows of beans or 2 rows of peas per bed every 15 days for continuous harvesting.

- **Plant onion and shallot bulbs.** In mild climates (wait until mid-March in harsher climates), loosen well-drained soil then make furrows every 6 inches (15 cm). According to an old French saying, the onions need to be able to hear the bells toll so plant at a shallow depth.

- **Start sowing garden cress, mesclun, and baby greens.** In loose, carefully leveled soil, broadcast seeds, then lightly bury them with a rake. Water regularly to keep the soil moist until germination. Harvest a few weeks later.

Planting Under Cover

- **Plant asparagus.** Choose a location in full sun with well-drained soil rich in organic matter. To prepare the plot, dig trenches 12 to 18 inches (30–45 cm) apart then plant asparagus crowns about 8 inches (20 cm) deep. Remember that this crop will occupy the bed for several years

Don't Forget

- **Check cold protection equipment.** February can still be a very cold month so walk through your vegetable garden, inspecting floating row covers and plastic films on tunnels, and pull them back into place as needed.

- **Buy pre-sprouted storage potatoes.** A few weeks before planting, allow potatoes to sprout in crates kept in a cool, bright space with temperatures between 50 °F and 60 °F (10–15 °C), ideal conditions for sprouting.

- **Remove mulches.** In colder climates and on empty plots, remove the mulch so the soil can gradually warm up. You might also want to cover them with silage tarp. In the following weeks, you can plant cold-resistant crops such as brassicas and onions.

- **Prepare tunnels.** Start preparing greenhouses and tunnels for the coming season. Check the condition of plastic films, ropes, and wires, clean the metal structures, and prepare the soil.

Harvests

You can continue to harvest winter vegetables that survived the cold:

- Spinach
- Mâche (corn salad)
- Chicory (escarole and curly endive)
- Kale
- Leeks
- Crosnes (Chinese artichoke), dug up on demand
- Salsify and scorzonera
- Jerusalem artichokes
- Radicchio di Verona and di Treviso
- Endives
- Winter lettuces
- Cabbages (below)

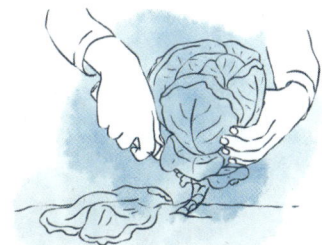

Tip from Jean-Martin Fortier

February, focus on growing quality seedlings, making sure you are providing the right conditions for them to thrive. Use high-quality seed starting mix, expose them to ideal temperatures for germination, and make sure they get enough light, otherwise they might become leggy. Always keep the soil moist but don't overwater. Seedlings that are well maintained from the outset produce strong, healthy plants, which are essential for you to have a successful year as a vegetable grower.

March

"As the days get longer and seedlings flourish, it's rewarding to look upon hundreds of thriving young plants. These early seedlings require careful monitoring and keep growers busy."
—Jean-Martin Fortier

Seeding Indoors

· **Sow tomatoes.** Fill open flats with potting mix then draw lines with a pencil and place a few seeds in each row. Make sure to sow thinly, being careful not to sow so many plants if your garden doesn't have enough space for them later. Leave seeded trays in a warm, bright space kept between 68 °F and 77°F (20–25°C).

· **Sow eggplants and peppers.** In an open flat filled with potting mix, make lines, drop seeds into them every inch or so, or broadcast seed thinly. Put trays in a bright space kept between 68°F and 77°F (20–25°C) and water regularly, keeping the soil moist until germination. Transplant seedlings into pots when they have 4 well-formed leaves.

· **Sow summer squashes and cucumbers.** Fill pots with potting mix, make 3 holes in each pot, and drop 1 seed per hole. This is sometimes referred to as cluster sowing. After germination, keep only the most vigorous plant from each pot to transplant in a greenhouse in April.

· **Seeding herbs (parsley, chives, marjoram, etc.).** Fill pots with potting mix, sow seeds, tamp the soil, water, then place in a warm, bright space.

· **Sow Swiss chard.** Get a jumpstart on the coming season with these leafy greens that can handle cool temperatures in the garden. Sow into open seed flats or a cold frame then transplant seedlings into the garden in rows 12 to 16 inches (30–40 cm) apart when they have 5 or 6 leaves. Cover with a row cover or low tunnel if a frost is forecasted.

Seeding Under Cover

- **Sow daikon radishes.** Under a cold frame or low tunnel, sow radishes in a shallow furrow then thin rows 3 weeks later when the seedlings are about 2 inches (5–6 cm) tall, keeping only 1 plant every 2 inches (5–6 cm).

- **Sow spring radishes.** Opt for 18-day varieties then broadcast sow and lightly cover the seeds. Water regularly but don't overwater so the radishes will be not too spicy and have a good crunch.

- **Sow beets.** Starting at the end of March, sow beets in rows 8 to 12 inches (20–30 cm) apart. Once they have germinated and developed 3 to 5 leaves, thin them by pulling plants that are overcrowded, keeping only 1 seedling every 4 to 5 inches (10–12 cm). You can replant the seedlings you pulled to create additional rows.

Direct Seeding

- **Sow turnips.** When the weather is right, loosen the beds, create a fine soil texture, and sow seeds in shallow furrows in rows 10 to 12 inches (25–30 cm) apart. Water to promote good germination then hoe and weed the bed if needed.

- **Sow peas and beans.** Carefully prepare the soil then sow seeds again to stagger harvests so you can enjoy early vegetables for longer. Hill young plants when they are 8 inches (20 cm) tall to help anchor the roots and keep the stems upright.

- **Sow parsnips.** Sow seeds in furrows about 2 inches (5 cm) deep and 8 to 10 inches (20–25 cm) apart. After germination, thin the rows, keeping 1 plant every 4 to 6 inches (10–15 cm).

- **Sow root parsley.** Choose a location with partial shade and loose well-drained soil rich in organic matter. Sow with the same spacing as parsnips. Water regularly to keep the soil moist until germination.

Field or Garden Planting

· **Transplant lettuces.** Loosen soil to create a fine texture and transplant seedlings (either purchased plugs or grown from seed) 10 to 12 inches (25–30 cm) apart in all directions. Remember that densely planted lettuces can be harvested like mesclun. Given more space, they produce nice heads of lettuce.

· **Transplant cabbages.** To ensure a cabbage harvest before summer, start by carefully preparing the soil and amending it with compost. Plant seedlings every 12 inches (30 cm) in rows 16 inches (40 cm) apart. Water thoroughly after planting and later only when the soil surface is dry.

· **Transplant artichokes.** Choose a sunny location with well-drained soil rich in organic matter. Plant seedlings 3 to 5 feet (1–1.5 m) apart and mulch with a layer of straw or compost to eliminate the need for weeding.

· **Planting *Helianthus strumosus* (woodland sunflower) and Jerusalem artichoke.** To get established and thrive in a vegetable garden, these perennial crops must be planted in rows 16 inches (40 cm) apart in loose soil amended with compost or organic fertilizer. But watch out because they can be invasive!

· **Pot up sweet peppers, hot peppers, and eggplants.** When the seedlings sown in February have 4 to 6 leaves, carefully dig them up and transplant individually into a 4- to 5-inch (10–12 cm) pot filled with potting mix. Tamp the soil and water. Keep seedlings in a tunnel or greenhouse until they are ready to go in the ground.

Don't Forget

- **Prepare beds for future crops.** If you covered the ground with silage tarps (occultation tarps) last fall to prepare them for future crops, it is time to remove them, loosen and amend the soil, and create beds to receive transplanted seedlings in the coming weeks.

- **Divide perennial herbs.** This is the time to divide perennial herbs such as mint, lemon balm, and chives to rejuvenate older plants and stimulate more vigorous growth.

- **Thin root vegetables.** Densely seeded carrots, turnips, radishes, or parsnips may need thinning to allow the remaining plants to grow better with more space to flourish.

- **Pre-sprout potatoes.** To ensure the earliest possible harvest, put seed potatoes in crates or egg cartons 1 month prior to planting, with the buds facing upwards to help them sprout. This may speed up the harvest by nearly 3 weeks.

- **Plant fruit trees, berry bushes, and mixed hedgerows.** The end of March is your last opportunity to plant these types of crops that often grow along the vegetable garden edges. Hedgerows can attract beneficial fauna (pollinating insects, birds), and fruit trees and berries (currants, blackberries, raspberries, etc.) provide harvests that complement your vegetable produce.

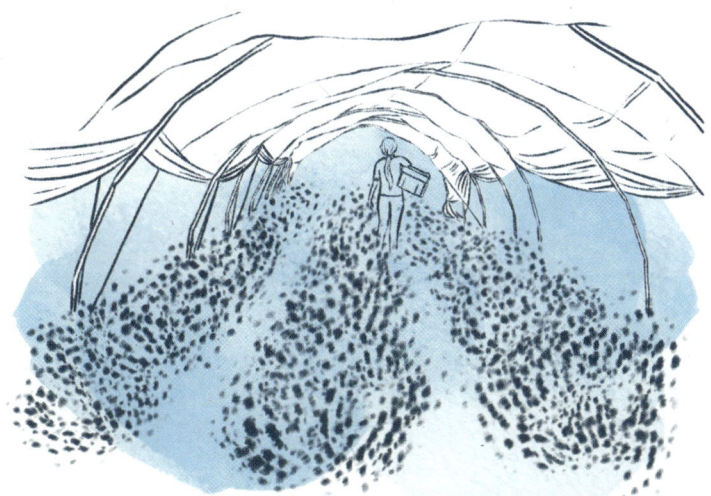

Harvests

With a tunnel or greenhouse, you can start harvesting vegetables in March that were sown at the beginning of the year:

- Radishes
- Lettuces
- Fennel (opposite)
- Swiss chard
- Spinach
- Mâche (corn salad)
- Chicory (escarole and curly endive)
- Kale
- Leeks

Tip from Jean-Martin Fortier

The weather in March can be unpredictable. Pay close attention to local forecasts and be ready to protect your plants with floating row cover if a cold snap or frost is expected. Also be mindful of longer sunny days that can quickly cause temperatures to spike in greenhouses and tunnels and harm young plants. Remember to get air circulating through these spaces to reduce the ambient temperature.

April

"In April, everything speeds up—seedlings planted under cover get bigger, and growers look for space to store them all. Outside field and garden crops take off, while maintenance tasks—weeding, hoeing, watering—start stacking up. But the days are getting longer too, and that's good for morale!"
—Jean-Martin Fortier

Seeding Indoors

· **Sow cucurbits.** Choose pots that are deep enough and fill them with potting mix rich in organic matter. Plant 2 zucchini, winter squash, or cucumber seeds per pot. Place them in a warm, bright space and water well to promote even germination.

· **Sow storage vegetables.** Start beets, celery, leeks, and brassicas in open flats, planting blocks, or plug flats kept in a warm, bright space. Transplant seedlings into the field or garden 1 month later. These vegetables keep well in a cellar so make sure you plant enough to last the winter—you wouldn't want to run out!

· **Sow annual herbs.** Prepare pots with a 50:50 mixture of potting mix and sand. Sow cilantro, dill, and especially basil seeds at a shallow depth. Tamp soil and water sparingly. Place them in a warm, bright room and keep the soil moist until germination.

Seeding Under Cover

· **Sow curly endive and escarole.** Sow 1 seed per cell in plug flats or planting blocks. Fill in the holes, tamp the soil, and water. Leave trays in a warm, bright space kept between 68°F and 72°F (20–22°C), under a cold frame or in a tunnel for planting in May.

· **Sow mesclun, Japanese mustard greens, and baby greens.** In loose, finely leveled soil, broadcast seeds then lightly rake the surface. Water with a gentle spray, keeping the soil moist until germination, then continue to water regularly. Harvest as needed.

Direct Seeding

- **Sow edible flowers.** Choose varieties such as nasturtiums, marigolds, and borage that add a splash of color to your dishes and offer unique flavors. Sow these flowers in a sunny location with loose soil.

- **Sow radishes, turnips, and beets.** Loosen the soil, dig a furrow with the handle of a tool, then sow thinly. Fill in the furrows, tamp the soil with the back of a rake, and water. After germination, thin the rows, keeping only 1 plant every 2 inches (3–5 cm). Regularly use a wire hoe to weed and loosen the soil.

- **Sow carrots, parsnips, and root parsley.** Loosen the soil and use a seeder to drop seeds every inch or so (2–3 cm) in rows 6 to 8 inches (15–20 cm) apart. Repeat the operation every month to stagger harvests. You can still sow parsnips and root parsley until the end of the month.

- **Sow chickpeas and lentils.** Loosen the soil and sow seeds roughly every 2 inches (4 cm) in rows 16 to 20 inches (40–50 cm) apart. Remember to stagger seedings for continuous harvests throughout the summer. Hill young plants when they are 8 inches (20 cm) tall to help anchor the roots and keep the stems upright.

Field or Garden Planting

- **Transplant brassicas.** It's time to transplant summer brassicas such as cauliflower, kale, and cabbage into loosened soil amended with compost or thoroughly composted manure. Cover the crops with insect netting to protect them from cabbage flies.

- **Plant hardy herbs (rosemary, thyme, sorrel, lemon balm).** Plant in a sunny location in well-drained soil. Water sparingly and harvest leaves a few weeks later to flavor your dishes.

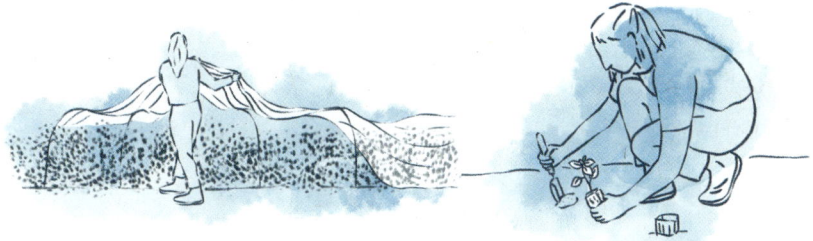

- **Plant potatoes.** Dig shallow trenches or holes in loose rich soil. Plant tubers that were pre-sprouted in April about 12 to 16 inches (30–40 cm) apart. Make sure sprouts (eyes) are pointing upwards then gently cover with soil. Water seedlings regularly and hill them when stems are 8 inches (20 cm) tall to encourage tuber growth.

- **Plant strawberries.** These grow successfully in a sunny location if the soil is carefully prepared and amended with organic matter (thoroughly composted manure or horn meal). Plant seedlings every 12 to 18 inches (30–45 cm) in rows 24 to 36 inches (60–90 cm) apart. Water regularly throughout the entire growing season.

- **Plant spring garlic.** In a sunny location, plant garlic cloves about 2 inches (5 cm) deep every 4 to 6 inches (10–15 cm) in loosened, well-drained soil. Rows should be 12 to 18 inches (30–45 cm) apart.

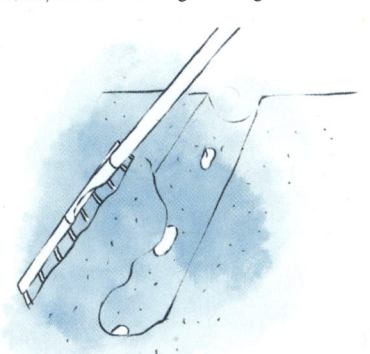

Crop Maintenance

· **Water and ventilate crops grown under cover.** April days can be quite sunny, making temperatures spike in cold frames, tunnels, and greenhouses. Open doors and raise the plastic film along the sides so fresh air flows through and lowers temperatures.

· **Weed frequently.** In the spring, you must keep the soil clean by eliminating weeds that are havens for slugs and other pests that compete with seedlings for nutrients. Using a collinear hoe to clear weeds also helps loosen the soil surface.

· **Trellis peas.** We recommend supporting peas as they develop to ensure more vertical growth and, especially, to make picking easier. Hill the plants and water often to keep the soil cool at all times.

Don't Forget

- **Carefully monitor fava bean seedlings.** Aphids can become a problem. If you notice an infestation, opt for natural control methods such as introducing ladybug larvae, applying nettle tea, or using an organic biocide.

- **Make nettle tea.** This is an excellent fertilizer and a natural plant treatment. Harvest fresh nettles, chop coarsely, put them in a large container, add water, and allow to ferment for a few days. Once the nettle tea is ready, dilute it with water before using it.

- **Harden off seedlings.** Gradually expose your seedlings to conditions outside the greenhouse or tunnel to help them acclimatize. Put them under a floating row cover that will filter sunlight then expose them to the sun for a few hours a day. Slowly increase the duration and intensity of exposure to sunlight and temperature changes. This process strengthens plant tissue and helps seedlings grow well after they are transplanted.

Harvests

The garden is starting to yield harvests (especially greens), and baskets are filling up!

- Spinach
- Lettuces
- Radishes
- Mâche (corn salad)
- Peas (early varieties, above)
- Green onions (scallions)
- Asparagus
- Kohlrabi
- Chives
- Rhubarb
- Sorrel

Tip from Jean-Martin Fortier

In April, be especially attentive to irrigation to ensure the success of your crops. Make sure to water regularly to avoid any dry spells and to shield seedlings from water stress. Consistently moist soil is essential for healthy plant growth, but it's equally important to avoid overwatering, which can make roots suffocate and rot. To get it right, we recommend gently poking your finger into the soil beside vegetables to monitor soil moisture. When it is dry to a depth of about 1 inch (2–3 cm), it's time to water.

May

"The weather is pleasant and warm in May, but watch out—depending on the region, this month can deliver a few surprise frosts. From mid-May (a period known in Europe as the Ice Saints), you can assume the risk of frost has passed and begin planting and sowing vegetable garden crops!"
—Jean-Martin Fortier

Seeding Under Cover

· **Sow pickling cucumbers.** Place a few seeds in pots filled with potting mix. Water and leave the containers in a bright location maintained between about 64°F and 70°F (18–21°C) then plant them in the garden when seedlings are about 4 inches (10 cm) tall.

· **Stagger lettuce, curly endive, and escarole seedings.** These crops usually mature at the same time, often resulting in harvest surpluses due to starting too many at once. You can stagger harvests by sowing or planting smaller amounts over a few weeks.

Direct Sowing
(as of mid-May)

- **Continue sowing storage vegetables.** Sow root vegetables such as turnips, beet, carrots, and parsnips about ¼ to ½ inch (1–2 cm) deep in furrows 10 to 12 inches (25–30 cm) apart.

- **Sow cucurbits.** Prepare the soil then drop squash and zucchini seeds into shallow holes in the ground. Water and cover with a cloche or row cover to speed up germination.

- **Sow bush beans.** In soil that has been loosened and finely leveled, sow seeds about 2 to 4 inches (5–10 cm) apart and keep the soil moist and cool until germination.

- **Sow sweet corn.** In a 2-inch (5 cm) deep furrow, drop corn kernels 12 to 14 inches (30–35 cm) apart then water. After germination, water and hoe the bed to eliminate weeds.

Field or Garden Planting

- **Plant storage potatoes.** Plant seed potatoes that have been pre-sprouted, by gently placing them, every 8 inches (20 cm) on the bottom of a 4- to 5-inch (10–12 cm) deep trench. Carefully fill in the trench to avoid damaging the sprouts.

- **Transplant summer brassicas (pointed cabbage, broccoli, cauliflower).** Loosen the soil with a broadfork then run a power harrow down the bed and mark out rows 16 to 20 inches (40–50 cm) apart. Plant seedlings, either purchased or sown in April, every 16 inches (40 cm). Tamp the soil, water, and cover with insect netting placed over metal hoops.

- **Continue planting lettuces.** For staggered consistent harvests, plant new seedlings (sown in April or purchased) every week to ensure fresh lettuces are available all summer.

- **Plant summer leeks.** Thoroughly loosen the soil and prepare leek seedlings, removing about a third of the foliage and lightly trimming the roots. Plant seedlings about 4 to 6 inches (10–15 cm) deep and 6 to 8 inches (15–20 cm) apart. Water often to help the plants get established.

- **Plant fruiting vegetables.** Plant vegetables such as tomatoes, peppers, and cucumbers in greenhouses or tunnels to protect them from the last frosts. Set up tomato stakes and beside each insert an upside-down plastic bottle with the lid removed and the bottom removed for watering.

Field or Garden Planting
(as of mid-May)

- **Transplant sweet peppers, hot peppers, and eggplants.** After loosening the soil with a broadfork and leveling it with a bed preparation rake, transplant seedlings every 24 to 28 inches (60–70 cm) in rows 32 inches (80 cm) apart. Trellis, water, and mulch.

- **Transplant tomatoes.** Choose a sunny spot and, if possible, a tunnel or greenhouse in cold climates. Plant tomatoes in rows spaced out enough to allow for air circulation then water. Set up stakes and loosely tie them to the stems.

- **Transplant summer squashes.** Grow in full sun and well-drained soil, leaving 24 to 36 inches (60–90 cm) between seedlings. Water regularly, mulch the soil, and watch out for slugs as they are fond of young summer squash leaves.

- **Transplant cucumbers.** Plant seedlings 30 to 36 inches (80–90 cm) apart and set up a support structure (mesh, tripod, bean pole) so that stems can cling to them and fruits can develop without touching the ground.

- **Transplant annual herbs.** Plant basil, cilantro, chervil, and dill in a warm, sunny location in well-drained soil rich in organic matter. Space plants about 10 to 12 inches apart (25–30 cm).

- **Transplant melons.** Dig holes that are slightly bigger than the root ball and transplant seedlings (purchased or sown in April) 24 to 36 inches (60–90 cm) apart. Water and mulch to keep the soil cool.

- **Transplant root vegetables.** Plant beet and celeriac seedlings sold in root balls with a dibber in loose soil amended with compost. Thoroughly water each seedling before and after planting.

Crop Maintenance

- **Monitor pests and diseases.** In May these threats appear from all sides. Closely examine leaves, young shoots, and stems so you can quickly spot any issues as they arise and take steps to protect your crops.

- **Mulch strawberry plants.** To prevent weed growth, maintain soil moisture, and avoid soil splash, mulch the crop with straw, dead leaves, or woven or nonwoven ground cover. Water often after the first flowers appear.

- **Mulch crops.** If your soil is warm enough and slugs aren't a problem, mulch with a good layer of straw, compost, or common reed to keep the soil cool and to suppress weed growth.

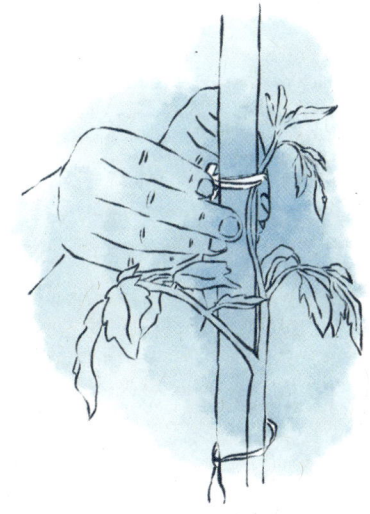

- **Thin seedlngs.** Root vegetables (carrots, parsnips, turnips, etc.) that were direct sown will need to be thinned. Remove overcrowded plants, leaving only the strongest ones spaced every 3 to 4 inches (8–10 cm).

- **Hill potatoes.** This is an important step when growing potatoes. Pull soil up to bury the base of the plants to promote tuber growth and protect them from sunlight.

- **Prevent diseases.** With disease-prone crops such as summer squashes, tomatoes, and fruit trees, spray the foliage with an undiluted horsetail solution (decoction). This preventive treatment will strengthen your plants.

- **Trellis and prune tomatoes.** Gently connect tomato stems to the lines or stakes you set up when transplanting seedlings and use this opportunity to remove suckers.

Harvests

At this time, early crop harvests are rolling in at a steady pace and summer vegetables are beginning to flower, promising bountiful harvests.

- Rhubarb
- Green peas
- Radishes (opposite)
- Baby carrots (above)
- Green onions (scallions)
- Sorrel
- Lettuces
- Asparagus
- Spring turnips
- Greenhouse beans
- Cauliflower
- Orach

Tip from Jean-Martin Fortier

May is a busy month with lots to do in the garden. Still take a moment to enjoy your garden: observe your growing vegetables, see their responses to weather events, and reflect on your horticultural practices implemented so far (mulching, watering methods, trellising or staking, etc.). And don't forget to appreciate the increased biodiversity in the garden.

June

"June provides a final opportunity to plant and sow vegetables you may have forgotten—if all the crop maintenance allows the time. It's one of the busiest months, so you must be well organized and, above all, get into the garden nearly every day."
—Jean-Martin Fortier

Seeding Under Cover

- **Sow summer varieties of curly endive and escarole.** Fill open flats or plug flats with seed starting mix then sow seeds ¼ to ½ inch deep (1 cm). Bury them, tamp the soil, and water with a gentle spray. Cover trays with plastic film to maintain a warm environment and leave them in a bright space. Keep temperatures between 59°F and 68°F (15–20°C) until germination then transplant seedlings when they have 3 or 4 leaves.

- **Sow fall brassicas (green and red cabbages, Brussels sprouts, kale).** There is still time to seed these crops in open flats or plug flats. Water and cover with insect netting.

- **Sow fennel, celeriac, celery, beets, and kohlrabi.** Sow in plug flats or planting blocks, cover with potting mix, and tamp the soil. Water with a gentle spray, place in a bright space kept at 68°F (20°C), and cover with insect netting.

- **Sow annual herbs.** Sowing herbs such as basil, parsley, dill, and cilantro will allow you to extend fresh harvests into late summer.

Direct Seeding

- **Sow summer squashes.** To guarantee fresh harvests throughout the summer, direct seed summer squashes in clusters. Water and provide some shade if it is quite hot during germination and development of the first leaves.

- **Sow pole beans and bush beans.** To ensure a summer harvest, we recommend sowing beans at the base of a support such as a steel trellis or a mesh panel hung on stakes.

- **Sow cut-and-come-again greens and arugula.** Choose oak leaf lettuce varieties to ensure a good summer mesclun harvest. Water regularly and shade young seedlings to keep them from going to seed.

Field or Garden Planting

- **Transplant summer vegetables.** You can still transplant zucchini, tomato, pepper, and eggplant seedlings into available bed spaces for late-season harvests. In loose soil, dig holes with a dibber, plant the seedlings, fill in, and tamp the soil. Set up stakes or trellis lines to secure the stems a few days later.

- **Replant lettuces.** This strategy maintains a constant supply of fresh greens. Transplant seedlings in shade so they can get established and water regularly to keep them from producing seed.

- **Transplant leeks, Swiss chard, and beets.** Plant seedlings (sold in planting blocks or plugs) in loose soil amended with compost. Water thoroughly, hoe often, and protect with a shade cloth, if needed, while they get established.

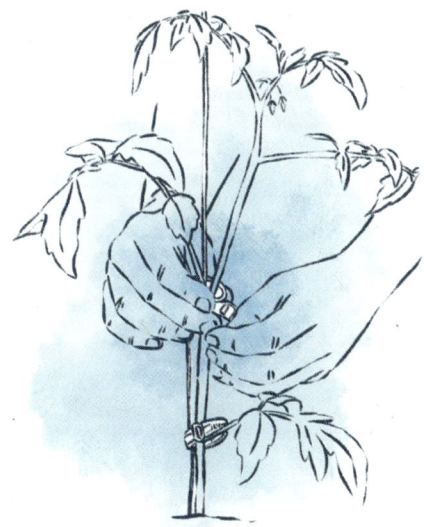

- **Prune eggplants, squashes, peppers, melons.** Pruning promotes plant growth and encourages production of higher-quality fruits. It also reduces the number of fruits so that the remaining ones grow bigger.

- **Apply liquid fertilizer.** To meet the nutritional demands of heavier feeders such as summer and winter squashes, tomatoes, peppers, and eggplants, we recommend applying a liquid fertilizer diluted in irrigation water. Doing this every 2 weeks promotes plant growth, fruit set, and ripening.

- **Hoe and mulch crops.** Carry out these tasks regularly to kill weeds, maintain loose soil, and retain moisture.

Crop Maintenance

- **Prune tomatoes.** Remove suckers, the shoots that form in the leaf axils along the main stem. This helps the plant direct energy towards fruit development rather than excess leaves and produces better yields.

- **Trellis tomatoes.** As tomato plants grow, gently tie the main stems to stakes or trellis lines, preventing them from sagging or collapsing under the weight of the fruits. Use string or, if growing under cover, plastic clips to attach stems to lines hung from the overhead structure in greenhouses or tunnels.

- **Monitor crops.** Crop monitoring is essential, allowing you to detect the appearance of pests, diseases, and nutritional deficiencies and quickly take action to prevent or address problems.

- **Remove rhubarb flowers.** We recommend removing the flower stems from the center of rhubarb plants to focus energy on producing edible stalks.

- **Trellis cucurbits.** Install stakes and supports to train the stems of climbing cucurbits such as cucumbers and pickling cukes to save space in a small garden, improve airflow, prevent disease, and facilitate harvesting.

- **Bury ollas.** These buried clay pots are an effective way to save water when irrigating, especially in arid regions, as they release water to plant roots and reduce water lost through evaporation.

- **Pinch basil.** Harvest basil by first pinching off the tips of stems to promote more compact bushy growth of bigger, more aromatic leaves.

- **Set up insect netting.** This is an effective way to protect crops from pest damage (cabbage, leek and carrot moths, and flies).

Harvests

Harvests are ramping up, as they now include both early vegetables and summer crops.

- New potatoes
- Baby carrots
- Broad beans (fava)
- Green peas, snap peas, snow peas
- Strawberries
- Swiss chard
- Turnips (opposite)
- Beets
- Summer squashes (bottom right)
- Spinach
- Onions and shallots
- Celeriac (below)
- Cucumbers
- Artichokes
- Sorrel
- Radish
- Cut salad greens, arugula, and mesclun
- Spring cabbages

Tip from Jean-Martin Fortier

The days are getting warmer and warmer, so it's best to harvest, plant, seed, and weed early in the morning or late afternoon. Soil maintenance (hoeing and hilling crops), pruning, and watering should be done in the evening when it's cooler. This approach is more comfortable for growers and better for plant health.

July

"In July microfarms and vegetable gardens are bursting with vegetables that are ready to be harvested, eaten, or canned."
—Jean-Martin Fortier

Seeding Under Cover

· **Sow Asian greens.** As with mesclun, sow mizuna, mibuna, and komatsuna into plug flats. Together, they make up a crisp and spicy salad mix. Water regularly and cover with insect netting to protect seedlings from flea beetles.

· **Sow winter radishes.** Sow into plug flats and place them in partial shade as winter radishes don't tolerate heat. They will be ready to harvest about 2 months later in the fall.

- **Sow kale.** Under a shade cloth, sow seeds roughly 1 inch (2–3 cm) apart and ¼ to ½ inch (1 cm) deep. Once seedlings are 4 to 6 inches (10–15 cm) tall, gently pull them out and transplant every 20 inches (50 cm) in rows 20 inches (50 cm) apart.

- **Sow Swiss chard.** Drop seeds every 2 to 4 inches (5–10 cm) into a furrow that is about ½ inch (1–2 cm) deep. Keep the soil moist until germination and, when the plants are roughly 4 inches (10 cm) tall, transplant them in loose, rich, cool soil.

Direct Seeding

- **Sow turnips.** At the end of the month, start sowing turnips again to ensure harvests in fall and winter. Direct sow into rows then cover lightly with soil. Water regularly to promote germination and thin when plants have 4 or 5 leaves.

- **Sow fall and winter greens.** Select cold-resistant varieties such as mâche (corn salad), chicory (sugarloaf, escarole), and winter lettuces.

- **Sow the last carrots.** Loosen the soil with a broadfork and carefully level the surface, sow in rows, cover with fine soil, and tamp with the back of a rake. Keep the soil cool and moist through regular watering and shade the seedlings if it is quite warm.

- **Sow fall and winter radishes.** To keep harvesting fresh radishes, sow winter varieties into carefully prepared garden beds. Set up metal hoops so you can lay insect netting over the seedlings. Towards the end of summer, replace it with a row cover to keep the crop warm.

- **Sow the last green beans.** Green beans need 60 days to reach maturity, so you still have time to sow for an October harvest. Sow in furrows 18 inches (45 cm) apart, dropping 1 seed every ¾ inch (2 cm). Fill in the furrow, tamp the soil, and water thoroughly.

Crop Maintenance

- **Fold or roll down onion tops.** This step enlarges the onion bulbs by directing energy away from foliar growth. Grasp the tops and push them towards the ground without breaking them.

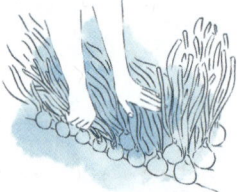

- **Pinch back winter squashes.** To develop larger fruits, pinch off the tips of fruit-bearing vines and remove stems without fruit to direct energy towards increasing the size of squashes.

- **Monitor caterpillars.** Watch out for caterpillars that can spread through brassicas and cause irreparable damage. Regularly inspect your plants and remove caterpillars by hand.

- **Monitor late blight.** In the summer, this fungal disease is more likely to arise on tomatoes, for example, when the weather is humid and mild. Get air circulating in tunnels and greenhouses, avoid overwatering (especially foliage), and preventively treat the foliage with a garlic or baking soda solution.

- **Water in the evening.** In the summer, we recommend watering at the end of the day when temperatures are cooler, which limits water loss through evaporation and allows vegetable crops to rehydrate more effectively.

- **Keep soil moist.** With recent plantings, keep the soil moist so they won't experience dry spells and water stress. Don't let the soil dry out completely between waterings.

- **Prune and trellis tomatoes.** To promote vertical growth, maintain good airflow around the base of plants, prevent disease, and facilitate harvesting, remove suckers and side shoots from tomato plants. Connect main stems to stakes or trellis lines.

- **Thin carrots.** Keep in-row spacing of 2 to 3 inches (5–7 cm) between plants to promote good root development. At this time, go over the bed with a flex tine weeder.

- **Prune and trellis eggplants, peppers, and cucumbers.** Cut back foliage that shades the fruit and pinch off the tips of fruit-bearing stems to promote ripening.

- **Hoe, mulch, and provide shade.** Use shade cloth, a shading mat made from wood slats, or reed fencing to protect heat-sensitive crops and seedlings from heat stress.

Don't Forget

- **Use silage tarps.** If you want to increase your garden space, lay a silage tarp over a grassy area covering the extent of area you wish to add. This destroys weeds and leaves soil that is ready for cultivation the following spring.

- **Ventilate greenhouses and tunnels.** Raise the side panels and open the doors of greenhouses and tunnels to avoid big temperature swings and overheating.

- **Turn over compost.** To speed up organic matter decomposition, stir compost heaps regularly, mixing the different layers to encourage decomposition of green waste. If the weather is dry, water the piles to maintain adequate moisture levels.

Harvests

Harvests have reached their zenith and feature an abundant diversity of vegetables that you can enjoy fresh or preserve in many different ways.

- Tomatoes (top right)
- Summer squashes
- Eggplants (middle right)
- Cucumbers
- Pickling cucumbers (to be harvested frequently)
- Carrots
- Onions
- Shallots
- Sweet peppers
- Beans (opposite page), broad beans, and peas
- Celery
- Turnips
- Fennel
- Spinach
- Artichokes
- Radishes
- Mesclun, arugula, baby greens (bottom right)
- Lettuces
- Strawberries
- Edible flowers (borage, nasturtium, marigold, etc.).

Tip from Jean-Martin Fortier

During heat waves, it's important to protect vegetable crops from high temperatures. Apply a thick mulch to retain soil moisture and reduce thermal radiation. Remember to water early in the morning or late in the evening to limit water loss from evaporation and irrigate at the base of plants rather than using sprinklers.

July

August

"Out in the field, it feels a bit like summer vacation, but August is a busy month for market gardeners and home gardeners alike. It's time to start planting fall and winter crops without neglecting crop maintenance and harvests...all while still taking time to enjoy yourself!"
—Jean-Martin Fortier

Seeding

· **Sow spinach.** Loosen the soil with a broadfork, incorporate compost, then use a bed preparation rake or garden claw to create a fine soil texture. Dig a furrow ¼ to ½ inch (1 cm) deep, sow seeds thinly, cover with soil, and tamp. Water regularly to keep the soil moist until germination.

· **Continue sowing chicory.** Fill an open flat with potting mix then sow and cover with ¼ to ½ inch (1 cm) more mix. Lightly tamp the soil and water with a gentle spray. Put the tray in a bright space kept between 60°F and 68°F (15–20°C). Once plants are about 2 inches (5–6 cm) tall, transplant them into the ground every 8 to 12 inches (20–30 cm).

Sow turnips. Prepare the soil by amending with compost and carefully smoothing the surface then sow seeds in shallow furrows 8 to 12 inches (20–30 cm) apart. Water regularly and use a shade cloth or a mat made from wood slats for protection as turnips do not tolerate much direct sunlight. After germination, thin the rows keeping 1 plant every 2 to 2½ inches (5–6 cm).

· **Sow mâche (corn salad).** Broadcast seed every 15 days until October to stagger harvests over several months.

· **Sow cover crops.** After pulling a crop from a bed (or turning it in), sow a cover crop to avoid leaving it bare. Instead of letting weeds grow, sow phacelia, mustard, vetch, or clover to improve soil fertility. In the fall, the stems and leaves of cover crops will be mowed and dug in to enrich the soil.

Field or Garden Planting

- **Transplant kale.** Plant multiple successions of kale seedlings into loosened soil until early October for harvesting until spring.

- **Plant strawberries.** Choose a sunny spot with well-drained soil. Amend beds with thoroughly composted manure or mature compost, dig holes about 4 inches (10 cm) deep and 12 inches (30 cm) apart, and plant seedlings. Water thoroughly then cover the soil with straw, compost, or felt.

- **Transplant brassicas.** Seedlings are sold in in garden centers. Plant cauliflower, Chinese vegetables (napa cabbage, bok choy), and winter cabbages every 24 inches (60 cm) in rows 24 to 28 inches (60–70 cm) apart. Choose a sunny location with rich loose soil and water regularly. Cover with insect netting pulled over metal hoops.

- **Plant fall lettuces and greens.** In the summer, plant seedlings (curly endive, escarole, and various lettuces that can be purchased from local growers) into loosened rich soil with a smooth and level surface. Water regularly and provide shade, if necessary, in the first few weeks.

Crop Maintenance

- **Protect winter squashes.** Slip tiles, thin wood pieces, or duckboards under developing fruits to keep the skin off the moist soil and prevent rot.

- **Water cucurbits.** Winter squashes, pumpkins, cucumbers, and summer squashes such as zucchini need lots of water. Build a little watering basin around each plant so water seeps in around the roots; avoid wetting the foliage to prevent cryptogamic (fungal and algal) diseases.

- **Remove strawberry runners.** With everbearing varieties, we recommend removing runners so each plant can focus on root development and fruit production.

- **Set up insect netting.** To protect carrots and brassicas from insect pests in the early stages after planting, lay insect netting over metal hoops. In the fall, lay a floating row cover over these supports.

- **Prune fruiting vegetables.** Tomatoes, eggplants, peppers, and cucumbers produce lush foliage, sometimes to the detriment of fruit development. Trim any excess that is shading fruits, remove suckers, and pinch off the ends of fruit-bearing stems to promote ripening.

- **Trellis plants.** Train the stems of fruiting vegetable plants, connecting them to structures (stakes or lines tied to overhead tunnel structures) that guide their vertical growth and support developing fruit.

- **Fill in mulches.** When mulching crops—with compost, reed, straw, etc.—it's important to cover the soil thoroughly and, if needed, add a second layer to maintain moisture and suppress weed growth.

- **Hill leeks and potatoes.** Use a hoe to pull soil up to bury the bases of the stems to promote whitening for leeks and tuber growth for potatoes. At this time, weed and hoe between the rows.

- **Monitor powdery mildew.** Always be on the lookout for powdery mildew as this fungal disease thrives in hot, dry conditions. Inspect the foliage in cucurbit crops and leafy greens and take preventive measures such as spraying with baking soda.

Don't Forget

- **Dry herbs.** If you have an abundance of herbs, this is the perfect time to dry them. Use a dehydrator or let them dry on racks in a dark ventilated room then store in airtight containers.

- **Sow cover crops.** Sow mustard, phacelia, or clover in an empty plot to help enhance soil quality, suppress weed growth, and loosen and improve the soil with their roots. This better prepares the plot for the following year's crops.

- **Preserve surplus harvests.** Canning your produce, especially tomatoes, is the perfect way to enjoy the fruits of your labor in the winter too! Make a tomato passata or ratatouille then freeze or can in sterilized jars.

- **Propagate herbs.** Take cuttings from thyme, rosemary, sage, savory, or marjoram by snipping off 4- to 6-inch (10–15 cm) stem fragments. Remove the lower leaves then transplant cuttings into a mixture of soil and sand. Water regularly—but not excessively—then replant any cuttings that have formed roots.

- **Prepare storage spaces.** It's time to get your cellar or storage spaces ready to receive harvests, especially root vegetables. The area should be clean, ventilated, dark, and kept above freezing temperatures to store potatoes, celeriac, parsnips, beets, and other vegetables that can't overwinter in the garden.

Harvests

Vegetable gardens are especially bountiful in the summer, delivering an abundance of veggies to eat fresh or preserve.

- Tomatoes (below)
- Summer squashes
- Eggplants (aubergines)
- Cucumbers
- Melons
- Onions, garlic, and shallots
- Sweet peppers
- Beans and peas
- Celery
- Beets
- Fennel
- Spinach
- Artichokes
- Radishes (opposite)
- Mesclun, arugula, baby greens
- Lettuces
- Everbearing strawberries

Tip from Jean-Martin Fortier

Beware of gardening on hot days or during a heat wave. It's best to garden in cooler times, either early in the morning or late in the day. The same goes for harvests. Plan to harvest fruiting vegetable crops, greens, and herbs, the most sensitive to high temperatures, early in the morning. Root vegetables can be harvested throughout the morning as they are more heat tolerant. In all cases, avoid working and harvesting at peak temperatures.

September

"For market gardeners and home gardeners, early fall is a very busy time. There's still a lot of harvesting to do, especially with first storage crops and summer vegetables that have to be preserved. Most importantly, in milder climates, it's also time to start winter crops and prepare plots for spring crops."
—Jean-Martin Fortier

Seeding

- **Sow arugula and Asian greens.** Seed into plug flats using a dial seed sower. When seedlings have 3 to 5 leaves, transplant them into a caterpillar tunnel or low tunnel to ensure harvests of fresh greens throughout fall and winter.

- **Sow spinach successions.** Use plug flats or direct sow spinach in rows 12 inches (30 cm) apart, always in small quantities and staggered until mid-October. Transplant from flats once seedlings have 3 or 4 healthy true leaves and use a low tunnel or row cover to protect them from the cold.

- **Sow turnips and fall and winter radishes.** Loosen the soil with a broadfork then create a fine soil texture with a garden claw or rake. Sow about ½ inch (1–2 cm) deep into furrows 10 to 12 inches (25–30 cm) apart. Tamp the soil and water until germination then regularly but not excessively.

- **Sow white onions.** Drop white onion seeds every 4 inches (10 cm) or so into a ¼- to ½-inch (1 cm) deep furrow. Water and keep the soil moist to ensure good germination. Onions will be harvested the following spring.

- **Stagger mâche (corn salad) seedings.** In loose soil amended with compost and broken down with a tilther, broadcast the mâche seed in rows. Lightly cover with soil, water, and lay a row cover over the beds to speed up germination.

- **Sow Chinese cabbages.** In plug flats filled with potting mix, sow 1 seed per cell at ¼ to ½ inch (1 cm) deep then water lightly. Put the trays in a bright location and keep the soil moist. Transplant seedlings into the ground once they are about 4 inches (10 cm) tall.

- **Continue sowing cover crops.** In empty plots, sow mustard, clover, or phacelia.

Field or Garden Planting

- **Transplant celery.** In loosened soil amended with compost, plant seedlings (purchased from a market gardener or nursery) 12 inches (30 cm) apart in all directions. Keep the soil moist, hoe the beds, and lightly hill the plants for tender white stalks.

- **Transplant Swiss chard.** This crop grows well in winter and can stay in the ground until spring. In soil rich in organic matter and in a sunny location, plant seedlings every 12 inches (30 cm) in rows 12 to 16 inches (30–40 cm) apart. Water regularly and protect with a row cover if nights are cold.

- **Continue transplanting kale.** Kale seedlings can be transplanted between rows of bok choy planted in August. Space plants about 12 inches (30 cm) apart in all directions to give room to grow. After you harvest the bok choy, the kale will keep growing by itself over the winter.

- **Transplant winter lettuces and escarole.** Loosen and level the soil then use a row marker to draw out rows and row spacings into a grid pattern. Plant seedlings where the lines intersect, roughly every 10 inches (25 cm). Plant successions for continuous harvesting. Water sparingly and hoe.

- **Transplant winter leeks.** Buy bare-root leek seedlings and plant them every 5 to 8 inches (12–20 cm) in rows 10 inches (25 cm) apart. Bury them about 6 inches (15 cm) deep for tender white stalks. Water and hoe regularly.

- **Transplant perennial herbs.** Now is the time to plant perennial herbs such as rosemary, sage, marjoram, and mint. In well-drained loose soil in a sunny and wind-sheltered location, plant seedlings, which are sold in pots at nurseries.

- **Transplant perennial vegetables.** Sorrel, artichokes, Egyptian walking onions, lovage, cutting celery, or Daubenton's kale are vegetables that grow back every year and can complement your seasonal produce. They are hardy, undemanding, and low maintenance.

Crop Maintenance

- **Monitor leek moths.** Be alert for leek moth activity such as damaged leaves, which can be the first sign of infestation. Cover plants with insect netting pulled over wire hoops.

- **Prune tomatoes.** Remove leaves, often yellow, from lower sections of plants, cutting them flush with the stem. This allows light and air to get through, reduces the likelihood of disease, and also promotes better ripening and earlier color change.

- **Harvest seeds.** If you want to produce your own seeds, harvest healthy ripe tomatoes and extract their seeds, dry them, and store in a dry location.

- **Soil maintenance.** Weed, hoe, and clean empty plots then spread organic matter (compost, manure) to improve the soil, which will loosen over the winter

Don't Forget

- **Remove diseased leaves.** In crops such as tomatoes, summer squashes, cucumbers, and winter squashes, regularly remove leaves affected by powdery mildew.

- **Prune foliage.** Towards the end of summer, eggplant, tomato, squash, and melon foliage may keep the sun from reaching the fruits. Remove leaves and significantly reduce watering to give them enough light to reach maturity.

Tip from Jean-Martin Fortier

While you can harvest your own seeds, especially with tomatoes as well as other vegetables and flowers, I prefer to leave this task to seed companies who have the expertise needed to guarantee the quality and consistency of the seed varieties I use. Plus I can choose different varieties based on demand or try new ones. I've found this allows me to focus on my crops, plant care, and harvests, but I understand that for some home gardeners it can be very rewarding to save and grow their own seeds.

Harvests

As the last summer vegetable harvests come to an end, storage vegetables reach maturity and are placed into winter storage.

- Artichokes
- Leeks
- Beans
- Swiss chard
- Carrots (opposite)
- Broccoli
- Eggplants (aubergine)
- Cucumbers
- Summer squashes
- Sweet peppers
- Hot peppers (bottom, left)
- Tomatoes (bottom, right)
- Chicory
- Mesclun and baby greens
- Parsnips
- Turnips and rutabagas
- Radishes
- Potatoes

October

"In October, it's important to protect crops from harsh winter conditions. Mini tunnels, cold frames, and floating row covers are valuable allies to fend off the cold and protect vegetables as temperatures drop. While this equipment can be quite costly, especially for home gardeners, growers who invest will be rewarded with fresh vegetable harvests in the dead of winter."
—Jean-Martin Fortier

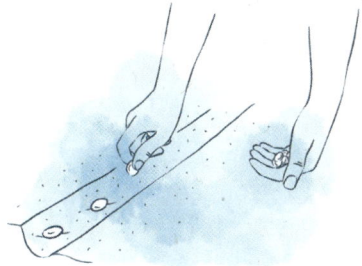

Direct Seeding

- **Sow spinach.** In loose soil, drop seeds every 6 inches (15 cm) into furrows about ¼ to ½ inch (1 cm) deep. Harvest the outer leaves first, taking care not to damage the center of the plant, to extend harvesting as long as possible.

- **Sow beans.** In mild climates, loosen the soil and amend it with compost, then sow seeds into a 2-inch (5 cm) deep furrow in rows 6 inches (15 cm) apart. Fill the furrows, tamp with the back of a rake, and water thoroughly. Protect the crop with a row cover or low tunnel if a cold winter is forecast.

Field or Garden Planting

- **Plant garlic.** Loosen the soil, add and incorporate compost, and smooth the surface. Make furrows about 2 inches (5 cm) deep then place a clove of garlic every 6 inches (15 cm). Fill in the furrow, tamp the soil, water thoroughly, and mulch with straw or leaves.

- **Transplant bok choy.** In loosened and leveled soil amended with compost or organic fertilizer (horn meal), plant seedlings every 8 to 12 inches (20–30 cm) in rows 12 to 16 inches (30–40 cm) apart. Water sparingly, hoe, and protect the crop with a floating row cover if a severe frost is forecasted.

- **Plant white onions and shallots.** Loosen the soil, level and smooth out the surface, then make furrows 8 inches (20 cm) apart. Plant seedlings that were sown in September, burying them ¾ inch (2 cm) deep and 4 inches (10 cm) apart. Water sparingly and, if temperatures drop below 32°F (0°C) or it is windy, protect the crop with a row cover or low tunnel.

Planting Under Cover

· **Sow arugula and claytonia.**
In unoccupied corners of your tunnel, sow rows of arugula and claytonia as they tolerate cold temperatures better than other greens. Water sparingly and protect the seedbed with a row cover until germination.

Crop Maintenance

· **Divide rhubarb.** For healthy, vigorous rhubarb plants, dig out the root ball, or crown, and cut it into pieces that have roots and at least 1 bud. Plant these in loose, rich soil then water once to settle the soil around the root balls.

· **Protect crops.** For cold-sensitive crops that will overwinter in the field, set up floating row covers, low tunnels, and cold frames to protect them from drastic temperature drops. Insulate the bottoms of brassicas and leeks with a layer of dead leaves or straw mulch.

· **Add compost.** To improve the soil, spread coarse compost and thoroughly composted manure over your plots, after mowing existing cover crops if necessary.

· **Harvest frost-sensitive vegetables.** Pick the last fruits before overnight temperatures get too low. You can then store squashes, sweet peppers, hot peppers, and tomatoes in a frost-free room where they continue ripening.

· **Bring endives indoors.** Using a broadfork, gently uproot the plants and store them in a dark frost-free cellar to encourage new shoots to sprout and yield white, tender endives.

- **Cut back bean plants.** After harvesting your bean crop, instead of pulling the plants, simply sever stems at the base, leaving the roots in the ground. Their nitrogen-rich nodules enrich the soil as they break down.

- **Cut back asparagus.** Remove any dry and yellowed stems, cutting them flush with the ground.

- **Blanch escarole.** Gather and tie leaves together with a string or rubber band to blanch the leaves at the center, making them more tender and less bitter. You can also cover them with an opaque plastic ground cover or terracotta cloche.

- **Water in the morning.** While evening waterings are highly recommended during hot weather, vegetables should ideally be watered in the mornings so the foliage and surrounding air dry out during the day. This approach reduces the likelihood of cryptogamic (fungal and algal) diseases.

- **Hill artichokes.** Bring soil or compost up around the base of plants to protect crowns from the cold.

Don't Forget

· **Pull out fruiting vegetables.**
In greenhouses and tunnels as well as outdoors, dig up tomato, eggplant, pepper, and cucumber plants. This frees up space to start winter crops. Hang tomato plants from the overhead structure in tunnels or greenhouses to allow the last fruits to ripen.

· **Collect green waste.** Stems, crop residues, and dead leaves are all organic matter in the making. Instead of throwing these out, use them as mulch on garden beds or compost them, especially tomato plants that could spread disease if used as mulch.

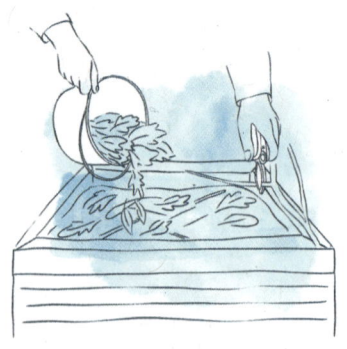

· **Clean trellising equipment.** Clean poles, stakes, and support structures in a tub filled with water and bleach. After they dry, coat with a Bordeaux mixture fungicide to prevent the spread of disease from one year to the next.

· **Aerate tunnels, greenhouses, and other shelters.** In dry mild weather, get air moving through tunnels, greenhouses, and cold frames regularly to reduce relative humidity and the likelihood of fungal disease.

Tip from Jean-Martin Fortier

Preserving garden vegetables allows you to eat fresh produce well beyond the summer. Take the time to research different food preservation methods: canning, freezing, dehydration, lacto-fermentation, and cellaring. You'll be able to enjoy the health benefits of your vegetables throughout the winter.

Harvests

When storing fall crops in a cellar or root clamp, regularly monitor their condition and remove any that are damaged.

- Winter squashes
- Winter radishes
- Beets
- Celeriac (below)
- Cardoon
- Swiss chard
- Cucumbers
- Tomatoes
- Sweet peppers, hot peppers
- Carrots
- Turnips and rutabagas
- Cabbages and kohlrabi
- Jerusalem artichokes
- Sweet potatoes (opposite)
- Parsnips
- Saffron

November

"Before winter comes, it's essential to take care of your vegetable garden, the crops still in the ground, and—especially—its soil. To help prevent erosion and maintain soil fertility in empty plots, cover beds with tarps or mulch, or seed with a cover crop that also prepares the soil for future crops."
—Jean-Martin Fortier

Planting Under Cover

- **Plant garlic, onions, and shallots.** Carefully prepare and loosen well-drained soil before planting bulbs purchased from a garden center. Opt to plant on a mound, into a 1-inch (2–3 cm) furrow, which helps the soil drain better. Place the bulbs or cloves in the furrow, fill it in, and water sparingly. If cold weather is forecasted, set up a floating row cover or low tunnel over the crop.

Seeding Under Cover

- **Sow parsley.** Loosen the soil with a tilther, level the surface with a rake, broadcast seed, and cover lightly with soil. Keep the soil moist and cover with a floating row cover until germination.

- **Sow spinach and mâche (corn salad).** In loose rich soil, sow in rows or broadcast seed. Lightly rake the surface to bury the seeds, tamp the soil, then water with a gentle spray.

- **Sow beans and peas.** Loosen and amend soil with compost then place seeds in furrows about 2 inches (5 cm) deep. Fill them in, tamp down, and water thoroughly. Watch out for rodents that are fond of these fleshy Fabaceae seeds.

- **Transplant brassicas.** In loose soil amended with compost, plant your last kale and cabbage seedlings. Make sure to have cold-protection equipment on hand such as cloches, floating row covers, and low tunnels.

Crop Maintenance

- **Protect crops.** Cover the soil and bottoms of leeks with straw or dead leaves to mitigate the effects of temperature fluctuations and allow you to pull them even after a hard frost. To protect artichokes from cold damage, cut back the stems then cover plants with dead leaves beneath a layer of soil.

- **Apply compost to perennial vegetables.** Spread a layer of compost around perennial vegetable plants to protect them from the cold and provide nutrients. Melting snow and rainwater will carry dissolved nutrients into the soil to feed the root systems.

- **Hoe fall crops.** In dry weather, use a stirrup hoe to kill weeds remaining among fall crops and loosen the soil surface compacted by autumn rainfall.

- **Monitor cold-protection equipment.** It's essential to carefully monitor your cold-protection equipment such as row covers, tunnels, and cloches. Check that they are in good shape, firmly anchored in the ground, and have no gaps that wind and cold air could enter.

November

ps. Keep a close eye
in your cellar or root
y are still in good
of decay, and above all
dents.

s (winter radish,
r after they've been
weather as it
ontent. Harvest them
sts!

ft)

ter)

overcome
to grow
necessarily
ed from

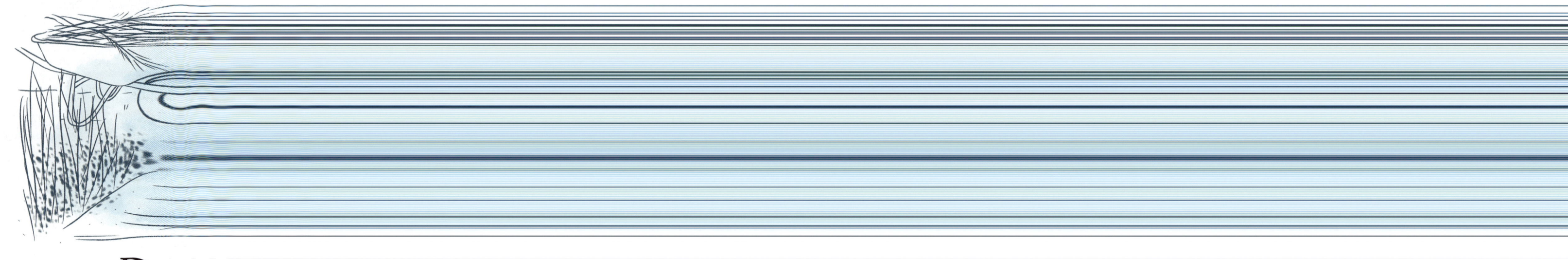

D

"In Quebec, winter c
there's even snow i
to harvest the leas
floating row covers
ground. Once that's
and surrounding ar
projects a

· **Monitor storage crops.** Keep a close eye on vegetables stored in your cellar or root clamp. Make sure they are still in good shape, show no signs of decay, and above all are protected from rodents.

Harvests

Certain root vegetables (winter radish, carrot, etc.) taste better after they've been exposed to some cold weather as it increases their sugar content. Harvest them right after the first frosts!

Don't Forget

· **Purge irrigation infrastructure.** It's important to be vigilant and take action to protect drip lines, sprinkler heads, pipes, and watering cans from frost damage. Make sure you drain and flush all your irrigation equipment at the appropriate time for your climate.

· **Bolster crop protection.** As temperatures drop, use 1 or more layers of row cover to better protect crops growing in tunnels. This provides additional insulation, maintaining a more temperate environment so vegetables won't be exposed to excessive temperature fluctuations.

· **De-sprout potatoes.** Gently remove sprouts (eyes) from stored tubers that develop in warm, dark spaces. This will ensure good-quality potatoes and extend their shelf life.

· Brussels sprouts
· Mâche (corn salad) (left)
· Brassicas
· Leeks
· Salsify (vegetable oyster)
· Scorzonera
· Winter radishes
· Carrots
· Parsnips
· Beets
· Turnips (below)

Tip from Jean-Martin Fortier

Growing vegetables in the winter is quite the challenge! You have to overcome lack of light, cold, and sometimes snow. I think it's still worth trying to grow them under cover in a space kept above freezing temperatures but not necessarily heated. Some species can withstand the cold, provided they are shielded from drastic freeze-thaw cycles with the right cold-protection equipment.

December

"In Quebec, winter can come very early, and sometimes there's even snow in October. Growers have to hurry to harvest the least hardy crops and urgently set up floating row covers to protect those remaining in the ground. Once that's done, it's time to tidy up the garden and surrounding area and start thinking about future projects and next year's calendar."
—Jean-Martin Fortier

December

Seeding Under Cover

· **Sow radishes.** If you want to harvest early radishes, sow them into a hotbed. This cultivation practice relies on decomposing manure to release heat that warms the soil in an enclosed space under a cold frame.

· **Sow early lettuces.** In a tunnel, loosen and level the soil, smooth the surface, then sow a winter lettuce variety that is suited to your winter conditions. Germination and growth are both slow, but you can speed them up by laying a floating row cover over the beds.

Crop Maintenance

· **Protect crops.** Lay a floating row cover over any leafy greens such as spinach or lettuces remaining in the field. Using straw or dead leaves, cover the base of leeks and brassica plants to protect them from frosts and keep the soil from freezing so you can harvest without using a spade.

Don't Forget

· **Clean up the garden.** Prune, shred, and compost stems of perennial vegetables such as artichoke and asparagus, foliage and flowers left from summer crops, and branches of shrubs, hedges, and perennials surrounding the garden.

· **Set up silage tarps.** To stop soil from becoming waterlogged over long winter months, cover plots with silage tarps that will also suppress weed growth and promote earthworm activity that naturally loosens the soil.

Harvests

Vegetables left in the ground (leeks, brassicas, etc.) should be harvested midday when plant tissue has thawed and the ground is loose enough to pull up the crop.

- Kale
- Winter lettuces
- Turnips
- Mesclun
- Cabbages and kohlrabi
- Parsley
- Mâche (corn salad)
- Spinach
- Jerusalem artichokes
- Leeks
- Salsify (vegetable oyster)
- Scorzonera (right)

Tip from Jean-Martin Fortier

As the year draws to a close, I strongly encourage you to stop, take a moment to look back on the last 12 months, and prepare for coming seasons. Go back to your notes to see what worked well, identify the top harvests, and determine which growing practices produced the best results. Don't forget to assess what went wrong too, remembering mistakes and problems or setbacks you encountered.

This end-of-season review is invaluable. It allows you to learn from your mistakes and build on your successes, helping you fine-tune next year's operations calendar and avoid repeating the same mistakes. Recordkeeping helps track how your garden evolves over the years, which is especially helpful when implementing crop rotations so you don't grow the same thing in the same spot two years in a row.

You should also reflect on what could be improved, which plots might be used more effectively, which cultivation techniques you might adjust, and which tools you should buy. After all, beginning next month, January, it will already be time to start your next season.

Acknowledgments from Jean-Martin Fortier

I wish to thank the entire team at the Market Gardener Institute for encouraging me to pursue my mission, every day. A big thank-you also goes out to the Growers & Co. team, who pushes me to come up with new types of equipment! I especially want to acknowledge my partner, Maude-Hélène Desroches, who is an exceptional market gardener and a dear friend!

Acknowledgments from New Society Publishers

We extend a great thanks to Delachaux et Niestlé, the French publisher, for working with us to publish this English edition. Further thanks to the New Society Publishers team for producing the book and especially to Laurie Bennett for her meticulous attention to technical details and high-quality translation into English.

Acknowledgments from Delachaux et Niestlé

A big thank-you to Jean-Martin Fortier and his team at the Market Gardener Institute for this wonderful collaboration.

Our heartfelt thanks go out to Pierre Nessmann for putting us in touch with Jean-Martin, for thoroughly editing this collection, and for being so generous with his time. For this book, we owe him so much. We also wish to thank Flore Avram, whose illustrations give this collection a beautiful, simple character; to Grégory Bricout for graphic design that cleverly reflects Jean-Martin Fortier's spirit; and to Sandrine Harbonnier and Sabine Kuentz for their work on the text.

Reference Books

The Market Gardener Masterclass, www.themarketgardener.com
The Market Gardener: A Successful Grower's Handbook for Small-Scale Organic Farming, New Society Publishers, 2014.
Winter Market Gardening: A Successful Grower's Handbook for Year-Round Harvests, New Society Publishers, 2023.
Microfarms: Organic Market Gardening on a Human Scale, New Society Publishers, 2024.

Grower's Guides from the Market Gardener

Tomatoes: A Grower's Guide
Vegetable Garden Tools: A Grower's Guide
Root Vegetables: A Grower's Guide
Living Soil: A Grower's Guide
The Well-Planned Vegetable Garden: A Grower's Guide
Fruiting Vegetables: A Grower's Guide

Coming Soon

Fall and Winter Vegetables
Starting and Propagating Plants
Salads and Leafy Greens
Herbs
Perennial Vegetables

Translator: **Laurie Bennett**

About New Society Publishers

New Society Publishers is an activist, solutions-oriented publisher focused on publishing books to build a more just and sustainable future. Our books offer tips, tools, and insights from leading experts in a wide range of areas.

We're proud to hold to the highest environmental and social standards of any publisher in North America. When you buy New Society books, you are part of the solution!

At New Society Publishers, we care deeply about *what* we publish — but also about *how* we do business.

- This book is printed on **100% post-consumer recycled paper**, processed chlorine-free, with low-VOC vegetable-based inks (since 2002)
- Our corporate structure is an innovative employee shareholder agreement, so we're one-third employee-owned (since 2015)
- We've created a Statement of Ethics (2021). The intent of this Statement is to act as a framework to guide our actions and facilitate feedback for continuous improvement of our work
- We're carbon-neutral (since 2006)
- We're certified as a B Corporation (since 2016)
- We're Signatories to the UN's Sustainable Development Goals (SDG) Publishers Compact (2020–2030, the Decade of Action)

To download our full catalog, sign up for our quarterly newsletter, and to learn more about New Society Publishers, please visit newsociety.com.

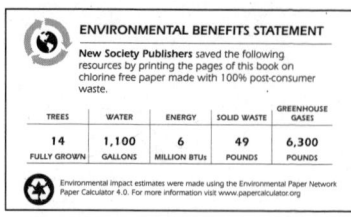

ENVIRONMENTAL BENEFITS STATEMENT

New Society Publishers saved the following resources by printing the pages of this book on chlorine free paper made with 100% post-consumer waste.

TREES	WATER	ENERGY	SOLID WASTE	GREENHOUSE GASES
14 FULLY GROWN	1,100 GALLONS	6 MILLION BTUs	49 POUNDS	6,300 POUNDS

Environmental impact estimates were made using the Environmental Paper Network Paper Calculator 4.0. For more information visit www.papercalculator.org

MIX
Paper | Supporting responsible forestry
FSC® C016245

www.newsociety.com